THE PAN-AFRICAN TELECOMMUNITY AND THE FUTURE OF DIGITAL INTEGRATION

Uwem Essia

AfCFTA: Pathways to the Largest Common Market in the World

Book 8

All rights reserved. No part of this book may be reproduced, distributed, or transmitted in any form or by any means, including photocopying, recording, or other electronic or mechanical methods, without the prior written permission of the publisher, except in the case of brief quotations embodied in critical reviews and certain other noncommercial uses permitted by copyright law.

Copyright © Uwem Essia October 2024, All rights reserved.

EXECUTIVE SUMMARY

"The Pan-African Telecommunity and the Future of Digital Integration" presents a comprehensive analysis and strategic framework for the urgent unification and advancement of telecommunications across Africa through the proposed establishment of a Pan-African Telecommunity (PAT). This research provides an in-depth exploration of the critical need for coordinated telecommunications development on the continent, examining historical contexts, current challenges, and future opportunities.

Historical Context and Current Landscape:

The study begins with a thorough examination of the historical evolution of telecommunications in Africa, tracing its roots from the fragmented colonial-era infrastructure to post-independence efforts at coordination. It critically analyzes the role of existing bodies like the African Telecommunications Union (ATU) in shaping the continent's digital landscape. A detailed assessment of the current state of telecommunications across African nations reveals persistent disparities in infrastructure development, regulatory frameworks, and digital access. The research identifies key challenges, including:

1. Uneven distribution of broadband access between urban and rural areas
2. Regulatory inconsistencies across nations

3. Infrastructure gaps and funding constraints
4. Technical and operational hurdles in cross-border connectivity

Rationale for the Pan-African Telecommunity:

Building on this analysis, the study presents a compelling case for establishing the PAT as a supranational body to address these challenges. Key arguments include:

1. The need for harmonized regulations to facilitate cross-border services and investments
2. The importance of coordinated spectrum management for efficient resource allocation
3. The potential for shared infrastructure development to reduce costs and expand coverage
4. The role of a unified body in representing African interests in global telecommunications forums

Organizational Framework and Governance:

The research proposes a detailed organizational structure for the PAT, including:

1. A General Assembly representing all member states
2. An Executive Council for policy implementation
3. Specialized committees focusing on critical areas such as spectrum management and cybersecurity

4. Regional offices to ensure effective implementation across the continent

A comprehensive governance framework is outlined, addressing:

1. Decision-making processes balancing consensus-building with efficiency
2. Mechanisms for ensuring fair representation of all member states
3. Strategies for engaging with industry stakeholders and civil society

Implementation Roadmap:

A phased implementation plan is presented, detailing:

1. Preparatory stages involving stakeholder consultations and legal framework development
2. Establishment and early operations focusing on pilot projects and capacity building
3. Expansion and consolidation phases for full-scale program implementation
4. Long-term strategies for establishing the PAT as a global leader in telecommunications policy

The research provides a detailed timeline with specific milestones and key performance indicators for each phase.

Addressing Challenges and Mitigation Strategies:

The study offers a nuanced analysis of potential obstacles, including:

1. Political and economic barriers such as sovereignty concerns and diverse policy priorities
2. Technical challenges related to interoperability and cybersecurity
3. Funding constraints and the need for sustainable financing models

For each challenge, the research proposes evidence-based mitigation strategies, drawing on case studies and best practices from successful regional integration efforts globally.

Future Prospects and Long-term Vision:

The research concludes with an exploration of prospects, focusing on:

1. The transformative potential of emerging technologies like 5G, IoT, and AI for Africa's economies
2. Strategies for positioning Africa in the global digital economy, including opportunities in digital services export, data centers, and e-commerce
3. The PAT's potential contributions to sustainable development goals in education, healthcare, and climate action

A comprehensive long-term vision is articulated, encompassing:

1. Universal broadband access targets
2. Digital skills development initiatives
3. Regulatory excellence and digital sovereignty

4. Sustainable and inclusive digital transformation across all sectors

TABLE OF CONTENTS

EXECUTIVE SUMMARY 2
Historical Context and Current Landscape: 2
Rationale for the Pan-African Telecommunity: 3
Organizational Framework and Governance: 3
Implementation Roadmap: 4
Addressing Challenges and Mitigation Strategies: 4
Future Prospects and Long-term Vision: 5
The Need for a Unified Approach to Telecommunications Development in Africa 20
Economic, Social, and Political Implications of Improved Telecommunications for the Continent 21
Purpose of the Book 22

1.1 The Colonial Genesis of Telecommunications in Africa ... 25
1.1.1 The Role of Telegraphy, Telephone, and Radio During the Colonial Era 26
1.1.2 Fragmented Infrastructure and Lack of Continental Integration 28

1.2 Post-Independence Telecommunications Landscape 29
1.2.1 Challenges Faced by African Nations 30
1.2.1.1 Infrastructure Gap 30
1.2.1.2 Technological Disparities 30
1.2.1.3 Funding Constraints 31

1.3 The Imperative for a Coordinated Telecommunications Approach 31
1.3.1 Telecommunications and Functioning in Modern Economies 32

1.4 Overview of the AfCFTA and African Vision 2063 34
1.4.1 The African Continental Free Trade Area (AfCFTA) 35
1.4.2 African Vision 2063 36
1.4.3 The Intersection of Telecommunications with AfCFTA and Vision 2063 36

2.1 Overview of Existing Infrastructure 39
2.1.1 Mobile Infrastructure 39

2.1.2 Fixed-Line Infrastructure 40
2.1.3 Broadband Infrastructure 40

2.2 Challenges and Regional Disparities 41
2.2.1 Urban-Rural Divide 41
2.2.2 Regional Disparities 41
2.2.3 Cost of Access 41
2.2.4 Regulatory Challenges 42
2.2.5 Power Infrastructure 42

2.3 Success Stories and Lessons Learned 42
2.3.1 Kenya's Mobile Money Revolution: M-Pesa 42
2.3.2 Rwanda's National Fiber Optic Backbone 43
2.3.3 Nigeria's Thriving Tech Hub 43
2.3.4 Morocco's Digital Development 43
2.3.5 South Africa's 5G Rollout 43

CHAPTER THREE: THE CASE FOR A PAN-AFRICA TELECOMMUNITY (PAT) 45

3.1 Benefits of a Unified Telecommunications Approach 45
3.1.1 Harmonized Regulations 45
3.1.2 Efficient Spectrum Management 46
3.1.3 Enhanced Cybersecurity 46
3.1.4 Improved Cross-Border Connectivity 47
3.1.5 Standardization of Infrastructure 47

3.2 Potential Economic and Social Impacts 47
3.2.1 Economic Growth 47
3.2.2 Job Creation 48
3.2.3 Digital Inclusion 48
3.2.4 Innovation and Entrepreneurship 49
3.2.5 Improved Public Services 49

3.3 Comparison with Other Regional Telecommunication Unions 49
3.3.1 European Conference of Postal and Telecommunications Administrations (CEPT) 49
3.3.2 Asia-Pacific Telecommunity (APT) 50
3.3.3 Arab Regulators Network of Telecommunications and Information Technologies (AREGNET) 50
3.3.4 Inter-American Telecommunication Commission (CITEL) ... 50

4.1 The Role of the Organization of African Unity (OAU) 53

4.2 The Founding Principles and Objectives of the ATU 54

4.3 Key Stakeholders and Leaders 55

4.4 Establishing the African Telecommunication Union (ATU) 56
    4.4.1 Structure, Governance, and Key Functions 57
    4.4.2 Legal Framework: Constitution and Bylaws 58

4.5 Vision of the ATU: Connecting Africa with Global Markets 58

CHAPTER FIVE: EARLY ACHIEVEMENTS AND CHALLENGES 60

5.1 Harmonizing Telecommunications Policies Across Member States 61

5.2 Capacity-Building Programs for Telecommunications Professionals 62

5.3 Challenges Faced 63
    5.3.1 Funding Issues and Reliance on International Donors 63
    5.3.2 Political and Economic Instability in Member Countries 64
    5.3.3 Disparities in Technological Development Among African Nations 65
    5.3.4 Internal Challenges: Bureaucratic Hurdles and Political Disagreements 65

6.1 Transformational Initiatives 67
    6.1.1 Mobile Telecommunications 67
    6.1.2 Satellite Communications 68
    6.1.3 Internet Connectivity 68

6.2 Development of Regional Frameworks 69
    6.2.1 Spectrum Allocation 69
    6.2.2 Internet Governance 70
    6.2.3 Cross-Border Connectivity 70

6.3 Impact on Economic Development and Regional Integration 71
    6.3.1 Telecommunications as a Driver of Economic Growth and Job Creation 71
    6.3.2 Promoting Cross-Border Trade and Regional Integration ... 72

7.1 Adapting to the Digital Revolution and New Telecommunications Realities 75
    7.1.1 Challenges of Internet Governance, Cybersecurity, and the Digital Divide 76
        7.1.1.1 Internet Governance 76
        7.1.1.2 Cybersecurity 76
        7.1.1.3 Digital Divide 77

7.2 ATU's Response to Digital Innovation 77
    7.2.1 Strategic Initiatives to Promote Digital Infrastructure and Services 77
        7.2.1.1 Infrastructure Development 77
        7.2.1.2 Standardization and Interoperability 78
        7.2.1.3 Support for Emerging Technologies 78

7.3 Collaboration with International Organizations and the Private Sector 79

CHAPTER EIGHT: KEY POLICIES AND FRAMEWORKS IMPLEMENTED BY THE AFRICAN TELECOMMUNICATIONS UNION (ATU) 81

8.1 Telecommunications Policy Harmonization 81
    8.1.1 Development of Unified Telecommunications Policies Across Member States 81
    8.1.2 Key Regulatory Frameworks Promoted by the ATU 82
        8.1.2.1 Spectrum Management 82
        8.1.2.2 Roaming Regulations 82
        8.1.2.3 Infrastructure Sharing 83

8.2 Digital Infrastructure Development 83
    8.2.1 ATU's Role in Promoting the Expansion of Broadband Infrastructure 83
    8.2.2 Efforts to Connect Underserved and Rural Areas 84
        8.2.2.1 Universal Service Funds (USFs) 84
        8.2.2.2 Community Networks 84
        8.2.2.3 Public-Private Partnerships 84

8.3 Capacity Building and Education 85
    8.3.1 Enhancing the Technical Expertise of Telecommunications Professionals in Africa 85
    8.3.2 Partnerships with Educational Institutions and International Bodies for Training and Knowledge Exchange 85

8.4 Collaborations with International Bodies 86
    8.4.1 International Telecommunication Union (ITU) 86
    8.4.2 World Bank 86
    8.4.3 African Development Bank (AfDB) 87

8.5 Collaboration for Funding 87
    8.5.1 Project Financing 87
    8.5.2 Capacity Building Programs 87
    8.5.3 Engagement of Multinational Telecommunications Companies 87

8.5.4 The Rise of Public-Private Partnerships (PPPs) in Telecommunications 88

CHAPTER NINE: ONGOING CHALLENGES AND FUTURE OUTLOOK 89

9.1 Ongoing Infrastructure Gaps and the Digital Divide 89
9.1.1 Regulatory and Policy Challenges in Adapting to New Technologies 90

9.2 Financial Constraints and Reliance on Foreign Aid 91

9.3 The Future of the ATU 91

9.4 The ATU's Vision for Africa's Telecommunications Sector in the Next Decade 92

9.5 How the ATU Aims to Address Emerging Challenges 93

CHAPTER TEN: THE BROADER IMPACT OF THE ATU ON AFRICAN SOCIETY 96

10.1: The Transformative Impact of Telecommunications on Education, Healthcare, and Governance 96
10.1.1 Education 96
10.1.2 Healthcare 97
10.1.3 Governance 97
10.1.4 Case Studies of Telecommunications' Impact on Quality of Life 98
    10.1.4.1 Kenya's M-Pesa 98
    10.1.4.2 South Africa's Education Initiatives 98
    10.1.4.3 Telemedicine in Nigeria 98

10.2 Empowerment Through Connectivity 99
10.2.1 Expanding Access 99
10.2.2 Digital Literacy Programs 99
10.2.3 Driving Youth Entrepreneurship, Innovation, and Social Mobility 100
    10.2.3.1 Youth Entrepreneurship 100
    10.2.3.2 Innovation Hubs 100
    10.2.3.3 Social Mobility 101

CHAPTER ELEVEN: THE BROADER IMPACT OF THE ATU ON AFRICAN SOCIETY 102

11.1 The Transformative Impact of Telecommunications on Human Development and Governance 102
10.1.1 Education 102

11.1.2 Healthcare 103
11.1.3 Governance 103
11.1.4 Case Studies of Telecommunications' Impact on Quality of Life 104
    11.1.4.1 Kenya's M-Pesa 104
    11.1.4.2 South Africa's Education Initiatives 104
    11.1.4.3 Telemedicine in Nigeria 105

11.2 Empowerment Through Connectivity 105
11.2.1 Expanding Access 106
11.2.2 Digital Literacy Programs 106
11.2.3 Driving Youth Entrepreneurship, Innovation, and Social Mobility 107
    11.2.3.1 Youth Entrepreneurship 107
    11.2.3.2 Innovation Hubs 107
    11.2.3.3 Social Mobility 107

CHAPTER TWELVE: STRUCTURE AND GOVERNANCE OF THE PROPOSED PAN-AFRICAN TELECOMMUNITY (PAT) 109

12.1 Organizational Framework 109
12.1.1 General Assembly 109
12.1.2 Executive Council 110
12.1.3 Secretariat 110
12.1.4 Specialized Committees 111
12.1.5 Regional Offices 111

12.2 Decision-Making Processes 112
12.2.1 Consensus-Based Approach 112
12.2.2 Tiered Decision-Making 112
12.2.3 Transparent Deliberation 113
12.2.4 Appeal Mechanism 113

12.3 Representation and Voting Rights 114
12.3.1 Universal Membership 114
12.3.2 One Country, One Vote 114
12.3.3 Weighted Voting for Certain Decisions 115
12.3.4 Regional Balance 115
12.3.5 Industry Representation 115
12.3.6 Quorum and Majority Requirements 116

CHAPTER THIRTEEN: KEY OBJECTIVES AND RESPONSIBILITIES OF THE PAN-AFRICAN TELL COMMUNITY (PAT) 117

    13.1 Standardization of Telecommunications Policies 117

13.1.1 Regulatory Harmonization: 117
13.1.2 Digital Market Integration: 118
13.1.3 Universal Service Policies: 118
13.1.4 Emerging Technology Regulation: 119

13.2 Spectrum Management and Allocation 119
13.2.1 Regional Spectrum Planning: 119
13.2.2 Harmonized Band Plans: 120
13.2.3 Spectrum Sharing Frameworks: 120
13.2.4 Digital Dividend Strategies: 120

13.3 Cybersecurity and Data Protection 121
13.3.1 Pan-African Cybersecurity Strategy: 121
13.3.2 Critical Infrastructure Protection: 121
13.3.3 Data Protection and Privacy Framework 122
13.3.4 Capacity Building in Cybersecurity: 122
13.3.5 Child Online Protection: 122

CHAPTER FOURTEEN: INFRASTRUCTURE DEVELOPMENT AND SHARING BY THE PROPOSED PAN-AFRICAN TELL COMMUNITY (PAT) 124

14.1 Strategies for Expanding Broadband Access 124
14.1.1 National Broadband Plans 124
14.1.2 Infrastructure Sharing 125
14.1.3 Public-Private Partnerships (PPPs) 125
14.1.4 Universal Service Funds 126
14.1.5 Innovative Technologies 126

14.2 Cross-Border Connectivity Initiatives 127
14.2.1 Regional Backbone Networks 127
14.2.2 One Network Area Initiatives 127
14.2.3 Internet Exchange Points (IXPs) 128
14.2.4 Harmonized Rights of Way 128

14.3 Shared Satellite and Undersea Cable Projects 128
14.3.1 Pan-African Satellite Initiatives 128
14.3.2 Regional Satellite Projects 129
14.3.3 Coordinated Undersea Cable Planning 129
14.3.4 Open Access Models 130
14.3.5 Capacity Building for Cable Management 130

CHAPTER FIFTEEN: REGULATORY HARMONIZATION OF THE PROPOSED PAT 132

15.1 Creating a Common Regulatory Framework 132
15.1.1 Model Regulations 132

- 15.1.2 Regulatory Impact Assessment 133
- 15.1.3 Capacity Building for Regulators 133
- 15.1.4 Regulatory Sandboxes 133

15.2 Addressing Licensing and Taxation Issues 134
- 15.2.1 Harmonized Licensing Framework 134
- 15.2.2 Regional Licensing 134
- 15.2.3 Tax Harmonization 134
- 15.2.4 Cross-Border Services 135

15.3 Ensuring Fair Competition and Market Access 135
- 15.3.1 Competition Policy 135
- 15.3.2 Infrastructure Sharing and Open Access 136
- 15.3.3 Number Portability 136
- 15.3.4 Interoperability Standards 136
- 15.3.5 Foreign Ownership Restrictions 137
- 15.3.6 Merger Control 137

CHAPTER SIXTEEN: DIGITAL INCLUSION AND SKILLS DEVELOPMENT UNDER THE PROPOSED PAT 139

16.1 Bridging the Digital Divide 139
- 16.1.1 Universal Service and Access Policies: 139
- 16.1.2 Community Networks 140
- 16.1.3 Public Access Points 140
- 16.1.4 Gender Digital Divide: 140
- 16.1.5 Accessibility for Persons with Disabilities: 140

16.2 Promoting Digital Literacy 141
- 16.2.1 National Digital Skills Strategies 141
- 16.2.2 Teacher Training 141
- 16.2.3 Online Learning Platforms: 141
- 16.2.4 Digital Skills for SMEs 142
- 16.2.5 Media and Information Literacy 142

16.3 Supporting Innovation and Entrepreneurship in the Telecom Sector 142
- 16.3.1 Innovation Hubs and Incubators: 142
- 16.3.2 Startup-Friendly Regulations 143
- 16.3.3 Intellectual Property Protection 143
- 16.3.4 Access to Finance 143
- 16.3.5 Digital Skills for Entrepreneurs 144
- 16.3.6 Regional and Global Market Access: 144

CHAPTER SEVENTEEN: FUNDING AND INVESTMENT STRATEGIES FOR THE PAT 146

17.1 Public-Private Partnerships (PPPs) 146
- 17.1.1 PPP Frameworks 146
- 17.1.2 Capacity Building for PPPs 147
- 17.1.3 Project Preparation Facilities 147
- 17.1.4 Regulatory Frameworks for PPPs 148

17.2 Regional Development Banks and International Funding 148
- 17.2.1 Collaboration with Regional Development Banks 148
- 17.2.2 Multilateral Development Banks 148
- 17.2.3 Bilateral Development Agencies 149
- 17.2.4 Climate Finance for Green Telecom 149

17.3 Innovative Financing Mechanisms 149
- 17.3.1 Infrastructure Bonds 150
- 17.3.2 Blended Finance 150
- 17.3.3 Impact Investing 150
- 17.3.4 Crowd-funding and Community Financing 151
- 17.3.5 Vendor Financing 151
- 17.3.6 Universal Service Funds 151
- 17.3.7 Spectrum Auctions and Licensing 152

CHAPTER EIGHTEEN: IMPLEMENTATION ROADMAP FOR REALIZING THE PAT 154

18.1 Phased Approach to Establishing the PAT 154
- Phase 1: Preparatory Stage (Year 1) 154
  - Stakeholder Engagement: 154
  - Legal Framework: 155
  - Institutional Design: 155
- Phase 2: Establishment and Early Operations (Years 2-3) 155
  - Ratification and Launch: 155
  - Staffing and Operations: 155
  - Initial Programs: 156
- Phase 3: Expansion and Consolidation (Years 4-5) 156
  - Full-scale Programs: 156
  - Partnerships: 156
  - Resource Mobilization: 156
- Phase 4: Maturity and Impact (Years 6 and Beyond) 157
  - Policy Influence: 157
  - Innovation Hub: 157
  - Digital Transformation: 157

18.2 Key Milestones and Timelines 157
- Year 1: 157
- Year 2: 158

Year 3: 158
Year 4: 158
Year 5: 158

18.3 Monitoring and Evaluation Mechanisms 159
18.3.1 Results-Based Management: 159
18.3.2 Annual Performance Reviews: 159
18.3.3 Independent Evaluations: 159
18.3.4 Stakeholder Feedback Mechanisms 160
18.3.5 Data-Driven Decision Making 160
18.3.6 Adaptive Management 160

CHAPTER NINETEEN: POTENTIAL CHALLENGES AND MITIGATION STRATEGIES 162

19.1 Political and Economic Barriers and Mitigation Strategies 162
19.1.1 Sovereignty Concerns 162
19.1.2 Diverse Policy Priorities 162
19.1.3 Economic Disparities 163
19.1.4 Funding Constraints 163

19.2 Technical and Operational Hurdles 164
19.2.1 Interoperability Challenges 164
19.2.2 Spectrum Coordination 164
19.2.3 Cybersecurity Threats 164
19.2.4 Skills and Capacity Gaps 165

19.3 Strategies for Building Consensus and Overcoming Obstacles 165
19.3.1 Inclusive Stakeholder Engagement 165
19.3.2 Evidence-Based Policy Making 165
19.3.3 Phased Implementation and Quick Wins 166
19.3.5 Flexibility and Adaptability 166
19.3.6 Leveraging Existing Initiatives 167

CHAPTER TWENTY: CASE STUDIES AND BEST PRACTICES 169

20.1 Lessons from Successful Regional Integration Efforts . 169
20.1.1 European Conference of Postal and Telecommunications Administrations (CEPT) 169
Key Lessons: 169
20.1.2 ASEAN Telecommunications Regulators' Council (ATRC) 170
Key Lessons: 170
20.1.3 West Africa Telecommunications Regulators Assembly (WATRA) 170
Key Lessons: 171

20.1.4 Eastern Caribbean Telecommunications Authority (ECTEL) 171
    Key Lessons: 171

20.2 Adapting Global Best Practices to the African Context 172
    20.2.1 Spectrum Management and Allocation 172
        Adaptable Practices: 172
    20.2.2 Digital Inclusion Initiatives 172
        Adaptable Practices: 173
    20.2.3 Regulatory Sandboxes 173
        Adaptable Practices: 173
    20.2.4 Infrastructure Sharing 174
        Adaptable Practices: 174
    20.2.5 Cybersecurity Frameworks 175
        Adaptable Practices: 175

CHAPTER TWENTY-ONE: FUTURE PROSPECTS AND LONG-TERM VISION 177

21.1 The Role of Emerging Technologies (5G, IoT, AI) 177
    21.1.1 5G Networks 177
        Key Considerations: 177
    21.1.2 Internet of Things (IoT) 178
        Key Considerations: 178
    21.1.3 Artificial Intelligence (AI) 179
        Key Considerations: 179

21.2 Positioning Africa in the Global Digital Economy 179
    21.2.1 Digital Services Export 179
        1.   Key Strategies: 180
    21.2.2 Data Centers and Cloud Services 180
        Key Considerations: 180
    21.2.3 Digital Financial Services 181
        Key Strategies: 181
    21.2.4 E-Commerce and Digital Trade 181
        Key Considerations: 182

21.3 Contribution to Sustainable Development Goals 182
    21.3.1 SDG 9: Industry, Innovation, and Infrastructure 182
    21.3.2 SDG 4: Quality Education 183
    21.3.3 SDG 3: Good Health and Well-being 183
    21.3.4 SDG 11: Sustainable Cities and Communities 184
    21.3.5 SDG 13: Climate Action 184

21.4 Long-term Vision 184

CHAPTER TWENTY-TWO: SUMMARY OF KEY FINDINGS, POLICY RECOMMENDATIONS, AND CONCLUSION 187

22.1 Key Findings 187
22.1.1 Historical Context and Current Landscape 187
22.1.2 Existing Coordination Efforts 187
22.1.3 Infrastructure and Investment 188
22.1.4 Emerging Technologies 188
22.1.5 Digital Economy 188
22.1.6 Sustainable Development: 189

22.2 Policy Recommendations 189
22.2.1 Establish the Pan-African Telecommunity (PAT) 189
22.2.2 Harmonize Regulatory Frameworks: 189
22.2.3 Coordinate Spectrum Management: 189
22.2.4 Promote Infrastructure Sharing and Development: 190
22.2.5 Foster Digital Innovation and Skills Development: 190
22.2.6 Enhance Cybersecurity and Data Protection: 190
22.2.7 Leverage Emerging Technologies: 190
22.2.8 Promote Digital Financial Services: 191
22.2.9 Support E-Commerce and Digital Trade: 191
22.2.10 Sustainable and Inclusive Development: 191

22.3 Conclusion 191

REFERENCES 193

INTRODUCTION

The history of telecommunications in Africa is deeply rooted in its colonial past. The introduction of telegraphy, telephone, and radio services during the late 19th and early 20th centuries was largely driven by European colonial powers. These systems, designed to serve the economic and administrative interests of the colonizers, resulted in a fragmented infrastructure that did not benefit the entire continent.

The development of telecommunications infrastructure during the colonial period was highly uneven. Advanced systems were concentrated in key colonial hubs, while vast rural areas remained disconnected. Telegraph lines, among the earliest forms of communication technology, were built mainly to enhance communication between colonial administrative centers, allowing swift decision-making during the "scramble for Africa." Egypt saw the installation of the first telegraph line in 1854, followed by similar efforts in South Africa and other British colonies.

By the mid-20th century, most African countries had rudimentary telephone services, though these were largely restricted to government officials and European settlers. The infrastructure, policies, and regulations were designed to meet the needs of the colonial powers rather than the local population. National systems were isolated from each other, and cross-border communication was both expensive and limited, reflecting a broader lack of regional integration within the telecommunications sector.

The Need for a Unified Approach to Telecommunications Development in Africa

Following independence, African nations faced enormous challenges in modernizing their telecommunications networks. A severe lack of infrastructure, coupled with insufficient funding and technological disparities, led to a fragmented telecommunications landscape. The need for a coordinated, unified approach to telecommunications development became evident as African countries sought to build robust systems that could drive socio-economic growth.

The absence of regional cooperation further exacerbated inefficiencies. Each country was left to independently develop its infrastructure, often resulting in isolated systems that failed to leverage economies of scale or share resources. This lack of collaboration hindered technological advancement and created gaps in service provision. However, with the advent of globalization and the increasing importance of telecommunications to the global economy, African nations recognized the pressing need for regional integration to remain competitive and spur economic development. A unified telecommunications strategy could facilitate cross-border trade, improve access to information, and drive economic growth by connecting isolated markets to regional and global networks.

Economic, Social, and Political Implications of Improved Telecommunications for the Continent

A cohesive telecommunications framework holds vast potential for Africa's economic, social, and political progress. Economically, telecommunications is a catalyst for productivity and innovation. Enhanced connectivity allows businesses to access larger markets, improves communication within and between countries, and boosts the digital economy. Research by the World Bank suggests that a 10% increase in broadband penetration could lead to a GDP growth of up to 1.38% in developing countries.

On a social level, improved telecommunications can help bridge the digital divide, connecting underserved and rural communities to essential services such as education, healthcare, and government programs. Access to telecommunications infrastructure promotes the dissemination of information and knowledge, fostering social inclusion and empowering marginalized groups.

Telecommunications is crucial for governance and public administration. It provides governments with efficient tools to communicate with citizens, enhances transparency and accountability by granting access to vital information and encourages citizen participation in the political process.

Purpose of the Book

To Provide a Comprehensive History and Analysis of the Establishment of the African Telecommunication Union (ATU)

This book aims to present a thorough historical account of the African Telecommunication Union's (ATU) formation. By exploring the conditions that led to its creation, the key stakeholders involved, and the challenges it faced, the book will offer an in-depth analysis of the role regional cooperation has played in shaping Africa's telecommunications landscape. Established in 1977, the ATU has been instrumental in coordinating telecommunications policies and strategies across the continent. Through its various initiatives, the ATU has fostered collaboration among African nations, working to harmonize telecommunications policies and standards.

To Explore the Impact of the ATU on African Telecommunications Policy and Development

The book will examine how the ATU has influenced telecommunications policy and development in Africa. It will delve into the role of the ATU in shaping regulatory frameworks, promoting infrastructure development, and encouraging cross-border cooperation. The ATU has played a vital part in facilitating the adoption of transformative technologies such as mobile telecommunications and broadband internet across the continent. By bringing together governments, private sector stakeholders, and international organizations, the ATU has provided a platform for the advancement of African telecommunications.

To Highlight the Ongoing Challenges and Opportunities in the Telecommunications Sector in Africa

While significant strides have been made, Africa's telecommunications sector continues to face numerous challenges, including infrastructure deficits, the digital divide, and regulatory hurdles. This book will explore these persistent issues in detail, providing the audience with a comprehensive understanding of the current state of the telecommunications sector in Africa. It will also highlight the vast opportunities present in the rapidly evolving digital landscape. With the rise of emerging technologies such as 5G, artificial intelligence, and the Internet of Things (IoT), Africa has a unique opportunity to leapfrog traditional development stages and become a global leader in telecommunications innovation. However, achieving this will require sustained investment, policy reform, and cross-border collaboration. By being aware of these challenges and opportunities, the audience can better prepare for the future of Africa's telecommunications sector.

To Propose Reforms Leading to the Formation of a Pan-African Telecommunity (PAT)

In addition to assessing the ATU's current efforts, the book will propose comprehensive reforms to unify telecommunications across Africa further. These reforms will outline a path towards the creation of a Pan-African Telecommunity (PAT), a vision of a seamlessly interconnected Africa where telecommunications serve as the backbone for economic growth, social inclusion, and political cohesion.

CHAPTER ONE: THE ORIGINS OF AFRICAN TELECOMMUNICATIONS

1.1 The Colonial Genesis of Telecommunications in Africa

The foundations of Africa's telecommunications infrastructure are inextricably linked to the continent's colonial past. In the late 19th century, European colonial powers introduced telecommunications technologies—such as the telegraph, telephone, and later radio—to Africa, primarily to serve their administrative, economic, and military objectives. The telegraph, one of the earliest forms of telecommunications, became a pivotal tool for colonial administrations to exert control over expansive territories, streamline governance, and maintain rapid communication with their European metropoles. This technology was not intended to promote local development but rather to facilitate the efficient exploitation of Africa's resources and consolidation of colonial power.

Each colonial power - be it British, French, Portuguese, or Belgian - developed its telecommunications infrastructure within its respective colonies, often neglecting the potential for interconnectivity between different colonial territories. For instance, the British established extensive telegraph networks in Egypt, Sudan, and South Africa, connecting these regions directly to London. However, these networks rarely extended to neighboring French or Portuguese colonies, resulting in a disjointed and fragmented telecommunications landscape across the continent. This segmented approach ensured that telecommunications infrastructure primarily served the interests of the colonizers, reinforcing colonial boundaries and inhibiting regional integration.

Telecommunications policies during the colonial era were meticulously crafted to align with the interests of the colonial powers. Colonial administrations maintained stringent control over telecommunications services, which were typically operated by state monopolies or European companies granted exclusive concessions. The high cost of telegraph and telephone services rendered them inaccessible to the majority of the African population. Consequently, telecommunications infrastructure was predominantly designed to cater to the colonial elite and European settlers rather than being developed as a public service to benefit the broader African populace. This instrumental view of telecommunications as a means to consolidate colonial dominance rather than as a tool for public empowerment deeply influenced the trajectory of telecommunications development in Africa.

1.1.1 The Role of Telegraphy, Telephone, and Radio During the Colonial Era

Telegraphy was the inaugural telecommunications technology introduced to Africa and played a crucial role in colonial administration. The first telegraph line in Africa was established in Egypt in 1854, connecting Cairo to Alexandria. Subsequently, telegraph networks expanded across North Africa, South Africa, and parts of West Africa. British and French colonial governments prioritized the development of telegraphy to link their African colonies with their respective European capitals. However, these telegraph networks were typically limited to major transportation routes such as railways and roads, which primarily served extractive industries and reinforced existing colonial infrastructure.

In the early 20th century, the telephone emerged as a vital communication tool in Africa. However, its usage was initially confined to colonial administrators and European settlers. Telephone networks were constructed in key urban centers like Nairobi, Lagos, and Accra, but access to telephone services remained scarce in rural areas where the majority of Africans resided. Telephone lines were predominantly used to connect government offices, businesses, and military installations, effectively excluding the wider population from accessing telecommunication services.

The introduction of radio in the 1920s and 1930s marked a significant development in African telecommunications. Unlike telegraphy and telephony, which were largely restricted to administrative use, radio broadcasting reached a broader audience and became a potent tool for shaping public opinion and disseminating colonial propaganda. Radio stations transmitted news, government announcements, and entertainment programs that reflected European interests, thereby extending colonial influence over local populations. Despite its limited initial reach, radio eventually became more accessible to the African public, particularly in the post-independence era, evolving into a powerful medium for political mobilization and social communication. However, during the colonial period, radio development was similarly controlled by colonial powers, with little emphasis on addressing the needs or aspirations of the African populace.

1.1.2 Fragmented Infrastructure and Lack of Continental Integration

The colonial approach to telecommunications development resulted in a highly fragmented infrastructure by the time African nations gained independence. Each colonial power developed its telecommunications systems tailored to its specific territories, leading to isolated networks that lacked interoperability. For example, British colonies such as Kenya, Nigeria, and South Africa had relatively advanced telecommunications systems compared to their French or Portuguese counterparts. However, these systems were often incompatible with each other, impeding regional communication and cooperation.

This fragmentation was further exacerbated by the absence of common standards or regulatory frameworks governing telecommunications across the continent. Each colonial power implemented its own policies and technological standards, resulting in significant disparities in the quality and extent of telecommunications services between different regions. The legacy of these fragmented infrastructures and divergent policies persisted well into the post-independence period, posing substantial barriers to regional cooperation and cross-border communication. The lack of continental integration in the telecommunications sector hindered the potential for a unified African network, limiting the continent's ability to leverage collective strengths and resources for comprehensive development.

1.2 Post-Independence Telecommunications Landscape

The wave of independence that swept across Africa in the 1960s and 1970s left newly sovereign nations inheriting outdated, fragmented, and ill-suited telecommunications systems for their burgeoning populations. National governments prioritized the modernization of telecommunications as part of broader efforts to develop national infrastructure and stimulate economic growth. However, the telecommunications systems left by the colonial powers were inadequate, presenting significant challenges to post-independence development.

Most African countries established state-owned telecommunications companies tasked with developing and managing national telecommunications networks. These entities were responsible for expanding telephone and radio services to previously underserved areas, particularly rural regions neglected during the colonial era. Despite these efforts, progress was often sluggish due to a multitude of obstacles, including the high costs associated with building telecommunications infrastructure, a scarcity of skilled technicians, and limited access to international funding.

1.2.1 Challenges Faced by African Nations
1.2.1.1 Infrastructure Gap

One of the most formidable challenges faced by African nations in the post-independence period was the significant infrastructure gap between urban and rural areas. Colonial governments had concentrated telecommunications development in major cities and administrative centers, leaving vast rural regions without basic services. This urban-rural divide continued to define telecommunications development post-independence as governments grappled with the complexities of extending services to remote and underserved communities. The lack of infrastructure in rural areas not only impeded communication but also limited access to vital services such as education, healthcare, and economic opportunities, thereby exacerbating social and economic inequalities.

1.2.1.2 Technological Disparities

Technological disparities further hindered the development of national telecommunications systems in Africa. Many newly independent countries were burdened with outdated equipment inherited from colonial administrations and lacked the financial resources necessary to upgrade or modernize their systems. The dependence on imported telecommunications technology from Europe and the United States made maintenance and expansion both costly and challenging. This reliance on foreign technology and expertise stifled the growth of local technological capabilities and innovation, contributing to the slow pace of telecommunications development across the continent.

1.2.1.3 Funding Constraints

Funding constraints posed another significant obstacle to telecommunications development in post-independence Africa. The construction and maintenance of telecommunications infrastructure required substantial capital investment, which many African governments struggled to secure. While international financial institutions such as the World Bank and the International Monetary Fund provided some assistance, their support often came with stringent conditions that limited the scope of government intervention in the telecommunications sector. These financial limitations restricted the ability of African nations to undertake large-scale telecommunications projects, further impeding the expansion and modernization of their national networks.

1.3 The Imperative for a Coordinated Telecommunications Approach

Recognizing the critical need for a more coordinated approach to telecommunications development, African nations began exploring regional cooperation in the 1960s and 1970s. Initial efforts focused on harmonizing telecommunications policies and standards to facilitate cross-border communication and promote economic integration. Key regional organizations, such as the Organization of African Unity (OAU) and the Economic Commission for Africa

(ECA), played instrumental roles in advocating for and fostering regional collaboration in telecommunications.

The establishment of the African Telecommunication Union (ATU) in 1977 marked a significant milestone in these efforts, representing the first formal attempt to create a unified framework for telecommunications development across the continent. The ATU was tasked with coordinating telecommunications policies, harmonizing technical standards, and promoting the sharing of resources and expertise among African nations. These early initiatives laid the groundwork for more comprehensive regional telecommunications strategies that emerged in the late 20th and early 21st centuries as African countries sought to integrate more fully into the global telecommunications landscape.

1.3.1 Telecommunications and Functioning in Modern Economies

In the rapidly evolving global economy of the 21st century, telecommunications has become a cornerstone of modern economies, serving as the essential infrastructure that connects individuals, businesses, and governments across vast distances. Telecommunications encompasses the transmission of information over significant distances through electronic means, and its impact extends far beyond mere communication. It is a key driver of productivity, innovation, and economic restructuring worldwide.

Numerous studies have highlighted the strong positive correlation between the development of telecommunications infrastructure and economic growth. For instance, a 2009 World Bank report found that a 10% increase in broadband penetration is associated with a 1.38% increase in GDP growth in developing countries. This relationship is particularly pronounced in Africa, where existing constraints in road, sea, and air transportation infrastructure limit economic mobility. Enhanced telecommunications connectivity has the potential to bypass these traditional development stages, enabling Africa to leapfrog into a more advanced and interconnected economic framework.

The transformative power of telecommunications is evident across various sectors:

1. Financial Services: Innovations such as mobile banking and digital payment systems have revolutionized access to financial services, especially in rural and underserved areas. Kenya's M-Pesa, which serves over 30 million users across East Africa, exemplifies how telecommunications can facilitate financial inclusion and economic participation.
2. Education: E-learning platforms and distance education programs are expanding access to quality education, bridging geographical and socioeconomic divides. These technologies enable students in remote areas to access educational resources and opportunities previously unavailable to them.
3. Healthcare: Telemedicine and health information systems are enhancing healthcare delivery, particularly in remote regions. These

technologies allow for remote consultations, efficient medical record management, and improved access to specialized medical expertise.
4. Agriculture: Mobile applications and Internet of Things (IoT) devices are transforming farming practices by providing real-time information on weather conditions, market prices, and pest control. These tools empower farmers to make informed decisions, increasing productivity and sustainability.
5. Governance: E-government initiatives are improving transparency, reducing corruption, and enhancing public service delivery. Digital platforms enable more efficient communication between governments and citizens, facilitating better governance and civic engagement.

1.4 Overview of the AfCFTA and African Vision 2063

To fully comprehend the significance of telecommunications in Africa's broader development trajectory, it is essential to examine how telecommunications interrelates with major continental initiatives such as the African Continental Free Trade Area (AfCFTA) and African Vision 2063. These initiatives provide a strategic framework within which telecommunications can significantly contribute to Africa's socioeconomic transformation.

1.4.1 The African Continental Free Trade Area (AfCFTA)

Implemented on January 1, 2021, the AfCFTA represents a monumental step toward continental economic integration, establishing the world's largest free trade area by the number of participating countries. The AfCFTA aims to create a single market for goods and services, facilitated by the free movement of persons, thereby deepening the economic integration of the African continent. Its key objectives include:

1. Creating a Single Continental Market for Goods and Services: By eliminating tariffs and non-tariff barriers, the AfCFTA seeks to enhance the flow of goods and services across African borders, promoting intra-African trade and economic collaboration.
2. Enhancing Competitiveness at the Industry and Enterprise Level: The AfCFTA aims to foster a competitive business environment that encourages innovation, efficiency, and growth among African industries and enterprises.
3. Facilitating the Movement of Capital and Natural Persons: By promoting the free movement of capital and people, the AfCFTA seeks to stimulate investment and labor mobility, contributing to economic dynamism and development.
4. Promoting Sustainable and Inclusive Socioeconomic Development: The AfCFTA emphasizes sustainable development practices and aims to ensure that economic growth

benefits all segments of society, fostering inclusive prosperity.

1.4.2 African Vision 2063
Complementing the AfCFTA, African Vision 2063 is the African Union's strategic framework for the socio-economic transformation of the continent over 50 years. Adopted in 2015, Vision 2063 envisions "an integrated, prosperous, and peaceful Africa, driven by its own citizens, representing a dynamic force in the international arena." Among its aspirations are:

1. A Prosperous Africa Based on Inclusive Growth and Sustainable Development: Vision 2063 aims to achieve sustained economic growth that benefits all Africans, ensuring environmental sustainability and social equity.
2. An Integrated Continent, Politically United, Based on the Ideals of Pan-Africanism: The vision underscores the importance of political unity and integration, promoting solidarity and cooperation among African nations.
3. An Africa of Good Governance, Democracy, Respect for Human Rights, Justice, and the Rule of Law: Vision 2063 prioritizes the establishment of robust governance structures that uphold democratic principles, human rights, and the rule of law.

1.4.3 The Intersection of Telecommunications with AfCFTA and Vision 2063
A viable Pan-African tele-connectivity framework stands at the confluence of these ambitious continental initiatives, playing a pivotal role in their realization. A unified telecommunications landscape

can significantly contribute to continental integration in several key ways:

1. Facilitating Trade: Seamless communication and data exchange are essential for the success of the AfCFTA. A harmonized telecommunications infrastructure can support e-commerce, digital trade platforms, and cross-border financial transactions, thereby enhancing the efficiency and reach of intra-African trade.
2. Enhancing Connectivity: Coordinated infrastructure development and spectrum allocation can bridge the digital divide between urban and rural areas, as well as between different regions of the continent. Improved connectivity ensures that all African populations can participate in and benefit from the digital economy.
3. Fostering Innovation: An integrated telecommunications market can drive innovation in telecom services and applications, attract investment, and nurture African tech startups. A robust telecommunications sector can catalyze technological advancements and entrepreneurial ventures across the continent.
4. Promoting Digital Skills: Unified standards for digital literacy and skills development can prepare Africa's workforce for the jobs of the future. Aligning telecommunications education with Vision 2063's emphasis on human capital development ensures that African populations are equipped to thrive in a digitalized economy.

5. Supporting Governance: Improved telecommunications infrastructure can enhance transparent governance, support democratic processes, and facilitate the implementation of pan-African policies. Digital platforms enable more efficient government-citizen interactions, fostering accountability and civic engagement.

By harnessing the power of unified telecommunications, Africa can accelerate its journey toward becoming a global powerhouse, realizing the vision of a connected, prosperous, and integrated continent. Telecommunications not only serves as the backbone for economic growth but also underpins social inclusion and political cohesion, making it an indispensable component of Africa's development strategy.

CHAPTER TWO: THE CURRENT STATE OF TELECOMMUNICATIONS IN AFRICA

The telecommunications sector in Africa, a topic of significant interest to academics, policymakers, and industry professionals, presents a dynamic yet uneven landscape. It is characterized by rapid advancements in certain areas while facing substantial challenges and disparities in others. This chapter delves into the current state of telecommunications infrastructure, explores regional and urban-rural divides, and highlights key success stories that provide valuable insights for future growth.

2.1 Overview of Existing Infrastructure

Over the past two decades, Africa has experienced remarkable progress in telecommunications, largely driven by the explosion of mobile technology. However, the development of fixed-line and broadband infrastructure lags, creating an imbalance in access and connectivity across the continent.

2.1.1 Mobile Infrastructure

Mobile technology is the cornerstone of telecommunications in Africa. By 2023, mobile penetration had reached 85%, translating to 1.08 billion mobile connections. The mobile sector has catalyzed access to communication, leading to:

1. Extensive 2G and 3G coverage across urban centers and rural regions.
2. Significant growth in 4G networks, especially in metropolitan areas.
3. Initial steps towards 5G implementation in countries such as South Africa, Kenya, and Nigeria signal a new era of connectivity.

2.1.2 Fixed-Line Infrastructure

Although mobile technology has revolutionized access to telecommunications, fixed-line networks remain essential for high-speed internet and robust business communications. However, fixed-line penetration remains limited, standing at just 0.9 per 100 inhabitants in 2022. This low figure highlights a significant gap in infrastructure development, particularly in rural and underserved areas.

2.1.3 Broadband Infrastructure

The expansion of broadband services is underway but faces significant hurdles. Despite these challenges, there are some bright spots:

1. Submarine Cables: By 2023, Africa was linked to over 40 submarine cables, dramatically enhancing international connectivity.
2. Terrestrial Fiber Networks: Around 1.5 million kilometers of terrestrial fiber will be laid by 2022, connecting key regions and boosting national data capacity.
3. Satellite Communications: In the remote areas where terrestrial solutions remain impractical, satellite communications, including new Low

Earth Orbit (LEO) constellations, promise to bridge connectivity gaps.

2.2 Challenges and Regional Disparities

While Africa's telecommunications landscape is evolving, it is marked by profound regional disparities, connectivity challenges, and infrastructure deficiencies. These obstacles must be addressed to unlock the full potential of digital transformation across the continent.

2.2.1 Urban-Rural Divide

A significant connectivity gap exists between urban and rural areas. While urban populations in Africa benefit from broad 4G coverage and relatively affordable data services, rural areas often struggle with even basic connectivity. In 2022, 81% of urban residents had access to 4G, compared to only 22% of the rural population.

2.2.2 Regional Disparities

Telecommunications infrastructure development varies widely across African regions. North Africa, for instance, boasts more advanced infrastructure and higher internet penetration. In contrast, Sub-Saharan Africa, particularly countries like Niger and Chad, struggles with internet penetration rates as low as 10%, further exacerbating regional inequalities.

2.2.3 Cost of Access

Despite efforts to lower data costs, internet access remains unaffordable for many Africans. In 2022, the average price of 1GB of mobile data represented 5.7% of an individual's monthly income—far above the UN

Broadband Commission's recommended affordability target of 2%.

2.2.4 Regulatory Challenges

The fragmented regulatory landscape across African nations presents a significant hurdle to cross-border connectivity and investment. Disparate policies complicate efforts to harmonize telecommunications standards, posing challenges to both operators and consumers. The need for regulatory harmonization is a pressing issue that must be addressed to unlock the full potential of digital transformation across the continent.

2.2.5 Power Infrastructure

Unreliable power supplies also hampered telecommunications services in Africa. Many regions face frequent power outages, which impact the reliability of telecommunications infrastructure and hinder consistent connectivity.

2.3 Success Stories and Lessons Learned

Despite the hurdles, several success stories across Africa offer valuable lessons for other countries seeking to enhance their telecommunications sectors.

2.3.1 Kenya's Mobile Money Revolution: M-Pesa

Kenya's M-Pesa, launched by Safaricom in 2007, has transformed financial inclusion not only in Kenya but also in other African markets. By 2023, M-Pesa had over 51 million active users across seven African countries. This success underscores the importance of regulatory flexibility and the value of partnerships between telecom operators and financial institutions.

2.3.2 Rwanda's National Fiber Optic Backbone

Rwanda's ambitious government-led initiative to build a national fiber optic network has paid off, with 95% of the country enjoying 4G coverage by 2023. This success underscores the potential for progress through public-private partnerships in infrastructure development and the necessity of long-term strategic planning.

2.3.3 Nigeria's Thriving Tech Hub

Nigeria, particularly Lagos, has emerged as a leading African tech hub, attracting substantial international investment. In 2022 alone, Nigerian startups raised over $1.2 billion in venture funding. This illustrates how robust telecommunications infrastructure can stimulate innovation, entrepreneurship, and economic growth.

2.3.4 Morocco's Digital Development

With 84% internet penetration by 2023, Morocco stands out for its comprehensive digital strategy. The government's proactive role in promoting digital adoption across sectors serves as a model for other African countries striving to achieve similar levels of digital development.

2.3.5 South Africa's 5G Rollout

South Africa leads the way in 5G deployment, with several major cities covered by 2023. Its success underscores the importance of timely spectrum allocation and highlights the potential of 5G to fuel industrial innovation and the development of smart cities, positioning South Africa as a leader in the global 5G race.

In summary, Africa's telecommunications sector stands at a critical juncture, offering both formidable challenges and immense opportunities. While the continent has made strides, particularly in mobile technology, there are still gaps to be filled in terms of broadband infrastructure, cost accessibility, and regulatory harmonization. However, success stories from countries like Kenya, Rwanda, and South Africa show that strategic investments and innovative approaches can drive rapid progress.

As the continent moves towards establishing an African Telecommunication Union, it must draw from these lessons to address infrastructure gaps and harmonize policies. With the right approach, Africa can unlock the full potential of a connected, digitally empowered future.

CHAPTER THREE: THE CASE FOR A PAN-AFRICA TELECOMMUNITY (PAT)

Africa is not just on the brink but in the midst of a digital transformation, with telecommunications playing a pivotal role in shaping the continent's future. The urgency of this transformation makes the establishment of a Pan-African Telecommunity (PAT) not just essential, but a pressing need for maximizing the benefits of this digital revolution. This chapter outlines the rationale for creating PAT, focusing on the advantages of a unified telecommunications framework, its potential economic and social impacts, and lessons from other successful regional telecommunication unions.

3.1 Benefits of a Unified Telecommunications Approach

A continent-wide telecommunications body like PAT offers numerous strategic advantages:

3.1.1 Harmonized Regulations

A PAT would streamline and harmonize telecommunications regulations across Africa. This would create a stable, consistent environment for investment, driving innovation and competition. By removing regulatory barriers between countries, PAT would encourage cross-border collaboration, spurring growth. The European Conference of Postal and

Telecommunications Administrations (CEPT), which has successfully aligned telecom policies across 48 European nations, is a model Africa could emulate. Harmonized policies also help create economies of scale, making it easier for new players to enter the market and offer competitive services.

3.1.2 Efficient Spectrum Management

Africa is entering a crucial phase of 5G deployment and the rise of Internet of Things (IoT) technologies. A PAT would enable more efficient, coordinated spectrum allocation and management, ensuring that this valuable and limited resource is optimized. This is vital as the demand for spectrum surges with the increasing adoption of advanced technologies. The Inter-American Telecommunication Commission (CITEL) has shown how regional coordination can lead to more effective spectrum use, setting an example for Africa.

3.1.3 Enhanced Cybersecurity

In today's interconnected world, cyberattacks are ever-present. A unified cybersecurity framework, supported by PAT, would bolster Africa's digital defences. By fostering collaboration on threat intelligence, joint assessments, and coordinated responses to cyber incidents, PAT could strengthen the continent's cybersecurity posture. The ASEAN-Japan Cybersecurity Capacity Building Centre offers a precedent for regional cooperation in tackling cyber threats, highlighting the potential benefits of such collaboration in Africa.

3.1.4 Improved Cross-Border Connectivity

A key advantage of PAT would be its ability to spearhead initiatives aimed at improving cross-border connectivity. This would significantly reduce communication costs and support initiatives like the African Continental Free Trade Area (AfCFTA). The West Africa Telecommunications Regulators Assembly (WATRA) has already made notable progress in this area, offering valuable lessons for PAT as it works towards a seamless, interconnected telecommunications network across Africa.

3.1.5 Standardization of Infrastructure

PAT could also play a vital role in promoting the standardization of telecommunications infrastructure across the continent. Standardized infrastructure would lead to better interoperability between countries and reduce costs associated with implementing disparate technologies. The European Telecommunications Standards Institute (ETSI) demonstrates how setting common standards can drive innovation and accelerate market growth, and this model can inspire similar progress across Africa.

3.2 Potential Economic and Social Impacts

The creation of PAT could trigger transformative economic and social changes across the continent:

3.2.1 Economic Growth

A PAT-led expansion in telecommunications infrastructure could not just accelerate, but significantly boost economic growth across Africa. Studies from the World Bank show that a 10% increase in broadband penetration in developing countries leads to a 1.38% increase in GDP growth. Through coordinated efforts to enhance broadband access, PAT could unleash tremendous economic potential, helping to narrow the digital divide and fostering greater economic activity in underserved regions.

3.2.2 Job Creation

The telecommunications sector, as a proven catalyst for job creation, offers exciting opportunities. The GSMA estimates that in 2022, the mobile ecosystem in Sub-Saharan Africa supported 3.8 million jobs. A PAT, with its potential to further drive job creation by facilitating market expansion and fostering a more dynamic, competitive telecommunications landscape, should be a source of excitement for professionals in the field. It could offer employment opportunities in both the private and public sectors, contributing to the growth and dynamism of the industry.

3.2.3 Digital Inclusion

A unified telecommunication strategy, a key focus of PAT, would aim to extend connectivity to the millions of Africans who remain unconnected. PAT's commitment to achieving the UN's Sustainable Development Goal (SDG) 9. c, which aims to significantly increase access to ICT and make internet access universal and affordable, should resonate with the audience's sense of social responsibility. By addressing the infrastructure needs of rural and underserved areas, PAT could promote digital inclusion, ensuring that no one is left behind in the digital age.

3.2.4 Innovation and Entrepreneurship

A harmonized telecommunications environment would nurture innovation and entrepreneurship across the continent. Improved connectivity, combined with streamlined regulations, would create fertile ground for tech startups and innovations to flourish. With cities like Nairobi, Lagos, and Cape Town already recognized as tech hubs, PAT could expand these success stories across the continent, fostering a new generation of African innovators.

3.2.5 Improved Public Services

The expansion and improvement of telecommunications infrastructure, under the guidance of PAT, could dramatically enhance the quality and availability of public services such as e-government, e-health, and e-education. These digital services would allow governments to provide more efficient, accessible, and transparent services to citizens, supporting broader development goals across Africa. The UN E-Government Survey 2022 emphasizes the transformative power of digital technologies in improving public service delivery.

3.3 Comparison with Other Regional Telecommunication Unions

A look at other successful regional telecommunication unions can offer valuable insights for PAT's design and implementation:

3.3.1 European Conference of Postal and Telecommunications Administrations (CEPT)

Since its inception in 1959, CEPT has been instrumental in harmonizing telecommunications policies across Europe. Its adaptable committee structure, focusing on electronic communications and radio spectrum, is a model that PAT could adopt. One key lesson from CEPT's experience is the importance of flexibility in accommodating rapidly evolving technologies while maintaining regional cohesion.

3.3.2 Asia-Pacific Telecommunity (APT)

Established in 1979, APT has played a crucial role in coordinating telecommunications development across the Asia-Pacific region. Its emphasis on capacity-building and technology transfer offers critical lessons for PAT, particularly in addressing the challenges of uneven infrastructure and limited expertise across Africa.

3.3.3 Arab Regulators Network of Telecommunications and Information Technologies (AREGNET)

Founded in 2003, AREGNET serves as a forum for Arab telecom regulators to coordinate policies and share best practices. Its initiatives, such as harmonizing roaming charges, provide insights into how PAT could address specific regional issues while aligning with global standards to promote greater integration.

3.3.4 Inter-American Telecommunication Commission (CITEL)

Operating under the Organization of American States, CITEL has successfully coordinated telecom policies across the Americas since 1994. Its working group structure, tackling issues such as spectrum management and digital technologies, could inform PAT's organizational framework, demonstrating how close coordination can support broader economic integration efforts.

In summary, the case for establishing a Pan-African Telecommunity (PAT) is both urgent and compelling. Drawing lessons from other successful regional unions while tailoring them to Africa's unique challenges and opportunities, PAT could serve as a cornerstone of Africa's digital future. The potential economic and social impacts—from GDP growth and job creation to enhanced public services and digital inclusion—underscore the importance of a unified telecommunications approach. As Africa embraces greater economic integration through initiatives like AfCFTA, the formation of PAT will not only be beneficial but essential for realizing the full potential of a connected, prosperous Africa.

CHAPTER FOUR: ORIGIN AND EVOLUTION OF THE AFRICAN TELECOMMUNICATION UNION

The inception of the Pan-African Telecommunity (PAT) builds on the existing African Telecommunication Union (ATU), which has a rich history rooted in Africa's post-independence movement. During this era, African nations recognized the need for a coordinated and unified approach to developing telecommunications. The fragmented infrastructure left behind by colonial rule posed significant challenges to cross-border communication. Each country operated on independent systems, creating a lack of interoperability that stifled political and economic cooperation across the continent.

The first substantial effort to address these issues emerged in the 1960s when the Economic Commission for Africa (ECA) initiated discussions on regional integration in the telecommunications sector. These early dialogues underscored the need to harmonize standards, develop shared infrastructure, and coordinate policies to enable seamless cross-border communication. The ECA's efforts were instrumental in laying the foundation for what would later become a continent-wide telecommunications body.

Regional and sub-regional initiatives also contributed to the vision of a pan-African telecommunications framework. For instance, the East African Posts and Telecommunications Corporation (EAPTC), created in 1963, unified postal and telecommunications services across Kenya, Tanzania, and Uganda, demonstrating the potential of regional cooperation. Other collaborations in West and Southern Africa further showed that shared telecommunications resources could foster connectivity and support economic development. These initiatives provided valuable insights into the benefits of cooperation, laying the groundwork for broader continental efforts.

4.1 The Role of the Organization of African Unity (OAU)

Established in 1963, the Organization of African Unity (OAU) played a pivotal role in the formation of the African Telecommunication Union (ATU). The OAU's recognition of telecommunications as a critical enabler of economic development and political integration led to its advocacy for a pan-African approach to overcoming the continent's fragmented telecommunications landscape. The OAU's call for African unity extended into crucial sectors like telecommunications, which were instrumental for facilitating trade, governance, and communication across Africa.

Responding to the OAU's advocacy, African governments and policymakers began considering the advantages of creating a continental telecommunications organization. The OAU played an instrumental role by convening high-level meetings that brought together telecommunications ministers, technology experts, and representatives from international organizations like the International Telecommunication Union (ITU). These discussions laid the foundation for a unified telecommunications body, with the OAU highlighting telecommunications as a tool for fostering African unity and identity.

By the mid-1970s, the OAU officially endorsed the establishment of a continental telecommunications union, recognizing it as essential for regional cooperation and integration. This endorsement provided the political momentum necessary for forming the ATU, an organization committed to fostering regional cooperation and integration. The ATU's establishment was a significant step towards achieving the OAU's broader goals of African unity and development.

4.2 The Founding Principles and Objectives of the ATU

In 1977, the African Telecommunication Union (ATU) was officially established with a clear set of guiding principles and objectives. The ATU's primary goal was to foster regional cooperation in the telecommunications sector, uniting African countries under a common framework to address the challenges of fragmented infrastructure and inconsistent national policies.

A key objective of the ATU was harmonizing telecommunications policies across the continent. This entailed developing common technical standards, regulatory frameworks, and spectrum allocation practices, ensuring that national systems were compatible with each other. Harmonization was crucial not only for cross-border communication but also for attracting investment, as a unified telecommunications market presented a more attractive opportunity to international investors and technology providers.

The ATU also focused on promoting the development of telecommunications infrastructure, particularly in underserved rural areas neglected during the colonial era. The organization encouraged knowledge sharing, capacity building, and collaboration among member states to bridge the infrastructure gap. This commitment to infrastructure development was not just about improving communication, but also about achieving broader development goals related to education, healthcare, and economic growth. The ATU recognized that a robust telecommunications infrastructure was a key enabler for these sectors, and it actively promoted its development to support overall socio-economic progress.

4.3 Key Stakeholders and Leaders

The establishment of the ATU was the result of efforts by multiple stakeholders, including African nations, industry leaders, and international organizations. Among the African countries leading the charge were Nigeria, Kenya, Egypt, and South Africa. These nations, which had relatively advanced telecommunications infrastructure at the time, recognized the strategic importance of regional cooperation and were instrumental in advocating for the formation of a continental telecommunications body.

Nigeria, in particular, emerged as a key proponent of the ATU, leveraging its status as one of Africa's largest economies and its strategic interest in regional leadership. Nigerian officials were actively involved in early discussions and hosted several key meetings that helped shape the ATU. Similarly, Kenya's experience in regional collaboration through the East African Posts and Telecommunications Corporation (EAPTC) provided valuable insights that influenced the ATU's vision and structure.

International organizations, especially the ITU, played a crucial role in the formation of the ATU. The ITU provided technical assistance, expertise, and financial support to African countries, ensuring that the ATU's framework aligned with global telecommunications standards. This alignment was crucial as it allowed African nations to integrate their systems with international networks, facilitating global communication and cooperation.

4.4 Establishing the African Telecommunication Union (ATU)

In December 1977, the ATU was formally established in Addis Ababa, Ethiopia, after years of deliberation and planning. This marked a major milestone in Africa's telecommunications history and represented a crucial step towards the regional integration of telecommunications systems. The creation of the ATU was driven by a collective recognition that the fragmented telecommunications infrastructure across the continent was a significant barrier to economic development. By providing a platform for African nations to collaborate on modernizing and integrating their systems, the ATU aimed to address these challenges and reduce reliance on external actors.

The OAU's endorsement of the ATU underscored the importance of telecommunications for achieving regional development goals. By establishing a dedicated body to oversee the sector, African leaders hoped to strengthen the continent's ability to develop its communications infrastructure and policies, furthering the objectives of economic growth and political unity.

4.4.1 Structure, Governance, and Key Functions

The ATU's structure was modelled after international telecommunications organizations like the ITU, ensuring it could effectively manage the diverse needs of its member states. The ATU's governance framework is built around three key bodies: the General Assembly, the Administrative Council, and the Secretariat.

The General Assembly, which consists of representatives from all member states, is the highest decision-making body of the ATU. It convenes periodically to set strategic priorities, approve policies, and elect members of the Administrative Council. The Administrative Council, composed of elected representatives, oversees the ATU's operations between General Assembly meetings and guides the Secretariat. The Secretariat, led by the Secretary-General, manages the ATU's day-to-day activities, coordinates programs, and represents the organization in international forums.

The ATU's core functions include promoting the development of telecommunications networks, standardizing telecommunications systems, and harmonizing policies among member states. The Union also provides technical assistance, facilitates capacity building, and promotes investment in the sector, playing a central role in driving the growth of telecommunications across Africa.

4.4.2 Legal Framework: Constitution and Bylaws

The ATU operates under a constitution and bylaws, which were adopted at its founding. The constitution outlines the Union's mandate, objectives, and governance structure, while the bylaws provide detailed guidance on operational matters. Together, they define the ATU's role in promoting regional cooperation, harmonizing telecommunications policies, and fostering telecommunications as a tool for economic and social development.

4.5 Vision of the ATU: Connecting Africa with Global Markets

The ATU's long-term vision is to position Africa as a global telecommunications hub. This vision involves improving both intra-African connectivity and the continent's links with global markets. By investing in high-capacity international gateways, undersea fibre-optic cables, and cutting-edge telecommunications technology, ATU seeks to integrate Africa into the global telecommunications landscape.

In line with this vision, the ATU also focuses on fostering innovation in areas such as mobile communications and broadband internet. By supporting research and development, the Union aims to promote technological self-reliance and reduce dependency on external technology providers. Furthermore, the ATU's vision encompasses the broader economic and social benefits of telecommunications, including poverty reduction, access to education and healthcare, and the promotion of economic growth across Africa.

CHAPTER FIVE: EARLY ACHIEVEMENTS AND CHALLENGES

During its formative years, the African Telecommunication Union (ATU) achieved significant milestones, laying the groundwork for its influential role in shaping Africa's telecommunications landscape. The ATU swiftly forged crucial partnerships and initiated several impactful projects aimed at fortifying the continent's telecommunications infrastructure. A key early success was the collaboration with the International Telecommunication Union (ITU), which provided vital technical expertise, financial support, and training programs. These initiatives were instrumental in enhancing telecommunication services across member countries, with a particular focus on connecting rural and underserved communities to both national and international networks.

A cornerstone project of the ATU's early efforts was the Africa Regional Communications Infrastructure Program (RCIP). This initiative aimed to enhance interconnectivity across the continent, reduce the cost of telecommunications services, and increase internet accessibility. By doing so, the ATU helped facilitate cross-border communications, which were crucial for regional economic integration and growth. The RCIP played a vital role in developing national backbone infrastructures, connecting African countries to undersea fibre-optic cables, and providing landlocked nations with much-needed access to the global internet.

Another significant partnership that bolstered the ATU's early work was with the United Nations Economic Commission for Africa (UNECA). Together, they embarked on developing regional telecommunications master plans aimed at creating standardized telecommunication systems and policies across the continent. These plans addressed key issues such as frequency management, equipment standardization, and establishing a common telecommunications regulatory framework. By fostering such uniformity, the ATU sought to encourage cross-border trade and investments while promoting technological cohesion across Africa.

5.1 Harmonizing Telecommunications Policies Across Member States

From its inception, the ATU's primary mission was to harmonize telecommunications policies across its member states. The Union aimed to foster collaboration between national regulators, governments, and telecommunications operators, aligning their policies with regional and international standards. To achieve this, the ATU organized numerous high-level forums, bringing together policymakers and industry leaders to discuss the adoption of uniform telecommunication regulations. These discussions were centered on reducing regulatory barriers to investment, simplifying licensing processes, and promoting fair competition within the telecommunications market.

A major milestone was reached in 1981 with the adoption of the African Regional Telecommunication Standardization Plan. This plan provided a blueprint for ensuring that African telecommunications networks were technically compatible and could interconnect seamlessly. The ATU also worked diligently on spectrum harmonization, ensuring that frequency bands allocated for telecommunications services were consistent across the continent. This alignment prevented interference and enabled more efficient network operations.

These harmonization efforts were instrumental in fostering a more integrated telecommunications market in Africa. By promoting greater interoperability between countries' systems, the ATU helped reduce operational costs for service providers, which in turn improved access to telecommunications services for consumers across the continent.

5.2 Capacity-Building Programs for Telecommunications Professionals

Recognizing the critical role of a skilled workforce in Africa's telecommunications sector, the ATU placed a strong emphasis on capacity-building from its early days. The Union partnered with international organizations, including the ITU and UNECA, to offer a range of training programs aimed at enhancing the technical capabilities of telecommunications professionals across Africa. These initiatives were designed to equip member countries with the expertise needed to manage and maintain their telecommunications infrastructure independently, reducing their reliance on foreign expertise.

One of the flagship programs launched in the ATU's early years was the Regional Training Program for Telecommunications Development. This program provided specialized training in areas such as network planning, frequency management, and telecommunications regulation. The curriculum also addressed emerging technologies like satellite communications and mobile telephony, which were anticipated to play transformative roles in Africa's telecommunications future.

To further support this goal, the ATU established several regional telecommunications training centers across Africa. These centers served as hubs for ongoing education, enabling professionals to upgrade their skills and share best practices across borders continuously. This approach not only strengthened the local workforce but also promoted greater collaboration and knowledge transfer between countries, fostering a sense of continental unity in the pursuit of telecommunications development.

5.3 Challenges Faced

5.3.1 Funding Issues and Reliance on International Donors

Despite its early achievements, the ATU encountered significant financial challenges. The Union's reliance on external funding from donors such as the ITU, UNECA, and various international development agencies meant that its financial sustainability was often in question. Many member states struggled to make consistent financial contributions due to domestic budget constraints, economic instability, and competing national priorities. This dependence on external support limited the ATU's ability to exercise full autonomy over its projects, leading to delays in implementation.

The financial shortfall also hindered the ATU's ability to embark on large-scale infrastructure projects. Although partnerships with international organizations helped bridge some of the funding gaps, it became clear that long-term sustainability would require increased financial commitment from member states. However, the diverse economic conditions across African countries complicated this effort, as wealthier nations were better positioned to contribute than their less developed counterparts.

5.3.2 Political and Economic Instability in Member Countries

Political and economic instability in several African nations further impeded the ATU's progress. Many member states were grappling with civil wars, political unrest, and economic crises, which diverted both attention and resources away from telecommunications development. In some cases, conflict led to the destruction of existing telecommunications infrastructure, setting back efforts to improve connectivity in affected regions.

Moreover, instability made it difficult for the ATU to foster regional cooperation. Despite these challenges, the ATU remained committed to its mission of promoting unity and collaboration across member states.

5.3.3 Disparities in Technological Development Among African Nations

A major challenge faced by the ATU was the wide gap in technological development between member states. While countries like South Africa and Egypt had relatively advanced telecommunications infrastructures, others lagged significantly behind, particularly in rural areas. This disparity complicated the ATU's efforts to promote uniform telecommunications policies and standards, as countries had differing capacities to implement these reforms.

The digital divide within Africa became increasingly evident as wealthier, more technologically advanced nations gained greater access to modern telecommunications services while poorer countries struggled to keep pace. The ATU recognized the need to address these imbalances through targeted investments and capacity-building initiatives, demonstrating its commitment to equality and progress.

5.3.4 Internal Challenges: Bureaucratic Hurdles and Political Disagreements

Internally, the ATU also faced bureaucratic obstacles that hindered its efficiency. Decision-making processes within the Union were often slow, as consensus among member states was required before any significant action could be taken. Political disagreements between countries, particularly over issues such as funding and resource allocation, further complicated the process, leading to delays and, in some cases, the scaling back of projects.

Despite these internal and external challenges, the ATU remained steadfast in its mission to improve telecommunications across Africa. It understood that overcoming these hurdles was crucial for the continent's long-term development and continued to work tirelessly toward its goals.

CHAPTER SIX: THE EVOLUTION OF TELECOMMUNICATIONS IN AFRICA

Since its inception, the African Telecommunication Union (ATU) has been a driving force in transforming Africa's telecommunications landscape. Recognizing the profound impact that modern telecommunication technologies could have on economic and social development, the ATU has spearheaded efforts to integrate cutting-edge technologies such as mobile telephony, satellite communications, and the internet, ensuring that these advancements reached all corners of the continent.

6.1 Transformational Initiatives

6.1.1 Mobile Telecommunications

One of the most transformative developments in African telecommunications has been the rise of mobile telephony, a revolution spearheaded by the ATU. In the late 1990s, the Union recognized that traditional landline infrastructure could not meet the continent's growing communication needs. In response, the ATU launched initiatives to promote mobile network expansion by encouraging governments to liberalize their telecommunications sectors. This liberalization allowed private investment and the entry of mobile operators into African markets.

The result was nothing short of revolutionary. According to the GSM Association (GSMA), mobile phone subscriptions in sub-Saharan Africa surged from a mere 16 million in 2000 to over 500 million by 2010. The ATU's advocacy for regulatory reform was crucial to this explosion in connectivity, as it pushed for flexible licensing regimes that spurred competition and innovation. The impact extended beyond communication: mobile telephony also paved the way for mobile banking and e-commerce, which have become essential pillars of Africa's economic growth.

6.1.2 Satellite Communications

Given the challenges of extending telecommunication infrastructure to rural and remote areas, the ATU championed satellite technology as a solution for bridging Africa's digital divide. Satellite communications provided the means to connect underserved regions that lacked access to terrestrial infrastructure. The ATU facilitated partnerships with international organizations to deploy satellite systems capable of delivering broadband services across vast, often isolated, areas. This technology enabled countries with limited infrastructure to offer their populations internet and telecommunication services, addressing disparities in access and ensuring that no region was left behind.

6.1.3 Internet Connectivity

The ATU's role in promoting internet adoption across Africa has been pivotal. It supported the development of regional Internet governance frameworks and initiatives such as the African Internet Exchange System (AXIS), which aimed to improve Internet connectivity by establishing Internet exchange points (IXPs) in member states. These IXPs allowed local internet traffic to be routed efficiently within Africa, reducing costs for internet service providers (ISPs) and making the internet more affordable for consumers. As a result, internet penetration in Africa grew exponentially, rising from less than 1% in the early 2000s to over 25% by 2020.

The efforts bore fruit, as internet penetration in Africa grew exponentially, rising from less than 1% in the early 2000s to over 25% by 2020. By fostering policies that encouraged broadband expansion and the growth of ISPs, the ATU contributed significantly to closing the digital divide and facilitating Africa's integration into the global digital economy.

6.2 Development of Regional Frameworks

The ATU's instrumental role in creating regional frameworks that govern spectrum allocation, internet governance, and cross-border connectivity is a testament to its efforts in fostering a cohesive and well-regulated telecommunications environment.

6.2.1 Spectrum Allocation

Efficient spectrum management is critical for the success of mobile and wireless services. The ATU has worked tirelessly to harmonize frequency allocations across its member states, ensuring that mobile communications operate on standardized frequency bands. This harmonization facilitates cross-border mobile services and helps telecommunications operators provide seamless service across regions, enhancing both connectivity and cooperation. By streamlining spectrum allocation, the ATU has played a central role in improving telecommunications efficiency and enabling better service delivery across Africa.

6.2.2 Internet Governance

As the internet became increasingly vital for Africa's economic and social development, the ATU took a proactive approach to shaping Internet governance. It facilitated discussions among key stakeholders, including governments, ISPs, and civil society, to create policies that promote internet access, cybersecurity, and network resilience. The ATU's collaboration with the African Union and other regional organizations has resulted in the establishment of cybersecurity frameworks to protect telecommunications infrastructure from emerging threats. This comprehensive approach to internet governance has strengthened Africa's capacity to safeguard its digital future.

6.2.3 Cross-Border Connectivity

The ATU's significant emphasis on improving cross-border connectivity is a testament to its vision of promoting regional integration and economic growth. By encouraging partnerships between neighboring countries, the Union has facilitated the development of transnational fiber-optic networks, such as the East African Marine System (TEAMS) and the West Africa Regional Communications Infrastructure Program (WARCIP). These initiatives have enhanced high-speed internet access, improved communication services, and fostered closer economic ties between African nations.

6.3 Impact on Economic Development and Regional Integration

The evolution of telecommunications in Africa has had far-reaching impacts on the continent's economic development and regional integration, with the ATU playing a key role in driving this progress.

6.3.1 Telecommunications as a Driver of Economic Growth and Job Creation

The growth of telecommunications infrastructure has been a major driver of economic growth in Africa. By creating an enabling environment for investment in telecommunications, the ATU has contributed to increased productivity, improved business operations, and job creation across various sectors. Mobile technologies, in particular, have opened up new opportunities for innovation and entrepreneurship.

One of the most notable examples is the rise of mobile money services, such as M-Pesa in Kenya, which revolutionized financial inclusion by providing millions of people with access to banking services. Mobile money's contribution to Kenya's GDP was estimated to be around 5%, highlighting the transformative power of telecommunications in driving economic progress. As more people gained access to communication and financial services, the telecommunications sector became a vital component of Africa's economic infrastructure.

6.3.2 Promoting Cross-Border Trade and Regional Integration

Telecommunications have also been critical to the success of cross-border trade and regional integration in Africa. Improved communication networks have enabled businesses to operate more efficiently across borders. At the same time, the ATU's efforts to harmonize regulations and promote standards have made it easier for companies to navigate different markets. These advancements have supported the growth of regional integration initiatives like the African Continental Free Trade Area (AfCFTA), which relies on robust telecommunications networks for seamless communication between trading partners.

In addition to reducing trade barriers, enhanced connectivity has driven down costs for consumers and businesses, fostering greater collaboration between African nations and contributing to the continent's overarching development goals. The ATU's focus on creating competitive telecommunications markets has been instrumental in promoting regional economic integration and ensuring that Africa remains on a path toward sustainable growth.

In summary, the ATU's initiatives in promoting mobile telecommunications, satellite communications, and internet connectivity have profoundly transformed Africa's telecommunications landscape. By establishing regional frameworks for spectrum management, internet governance, and cross-border connectivity, the Union has fostered greater regional integration and economic development. As Africa continues to embrace new technologies, the ATU's contributions will remain crucial to unlocking the continent's full potential in the digital age.

CHAPTER SEVEN: THE DIGITAL REVOLUTION AND ITS IMPACT ON THE ATU

The transformative power of mobile telecommunications has been a seismic force shaping Africa's digital landscape. Since the early 2000s, the continent has witnessed a monumental surge in mobile phone usage, with subscriptions surpassing 1 billion by 2021. This dramatic increase, a testament to the growing accessibility of mobile technology and the widespread rollout of mobile network infrastructure, has bridged communication gaps across both urban and rural areas, fundamentally reshaping how people and businesses interact.

According to the International Telecommunication Union (ITU), Africa's mobile penetration rate reached 100% by 2020, meaning that there were as many mobile subscriptions as people on the continent. This boom in mobile connectivity has revolutionized communication, facilitating everything from transactions to information sharing. The introduction of services like mobile banking, e-health, and e-learning has also greatly improved service delivery, enhancing the quality of life for millions of Africans.

The internet has similarly experienced rapid growth. The number of internet users in Africa grew from 4 million in 2000 to over 525 million by 2020, representing about 43% of the population. This expansion has been driven largely by mobile broadband services, which allow users to access the internet on smartphones. Additionally, the rise of social media, online marketplaces, and digital content has fueled internet usage, positioning Africa as one of the fastest-growing digital markets in the world.

7.1 Adapting to the Digital Revolution and New Telecommunications Realities

In response to these seismic shifts, the African Telecommunication Union (ATU) has proactively adjusted its strategies to align with the rapidly changing telecommunications environment. Recognizing the profound impact of mobile and internet technologies on socio-economic development, the ATU has emphasized the need for an enabling regulatory framework that fosters digital innovation.

The ATU's proactive stance is evident in its key initiatives, particularly in advocating for policies that encourage investment in digital infrastructure and services. The Union has worked closely with African governments to harmonize regulations, ensuring that digital services can be provided seamlessly across borders. This regulatory harmonization has been essential in creating a cohesive digital ecosystem across the continent.

Moreover, the ATU has emphasized the importance of developing digital literacy and skills within member states. Through various capacity-building programs, the Union has sought to equip individuals with the knowledge and skills needed to navigate the digital world effectively. This approach ensures that Africa's populations can fully leverage the benefits of the digital revolution.

7.1.1 Challenges of Internet Governance, Cybersecurity, and the Digital Divide

Despite the significant strides in mobile and internet adoption, Africa still faces critical challenges, particularly in the areas of internet governance, Cybersecurity, and the digital divide.

7.1.1.1 Internet Governance

The rapid expansion of Internet services has raised complex questions about governance, including data privacy, content regulation, and user rights. The ATU has been at the forefront of advocating for Internet governance frameworks that balance the need for free and open access with concerns over safety, security, and privacy. The Union has actively participated in international discussions aimed at developing policies that protect users while fostering innovation and growth in the digital space. This advocacy involves [specific activities or initiatives], demonstrating the Union's commitment to shaping the digital landscape in Africa.

7.1.1.2 Cybersecurity

As Africa's digital landscape expands, so do cybersecurity threats. Cyberattacks targeting businesses, governments, and individuals have become increasingly sophisticated and frequent. To address these risks, the ATU has promoted the development of comprehensive cybersecurity policies and strategies across member states. The Union has worked closely with international organizations and cybersecurity experts to share best practices and bolster the continent's defense against cyber threats.

7.1.1.3 Digital Divide

While mobile technology and internet access have spread across the continent, significant disparities remain, particularly between urban and rural areas. The ATU has recognized this issue and made it a priority to address it through infrastructure development and public-private partnerships aimed at improving access in underserved regions. The Union's commitment to inclusivity is evident in these efforts to bridge the digital divide.

7.2 ATU's Response to Digital Innovation

The ATU has responded to these challenges and opportunities by implementing several strategic initiatives aimed at promoting digital infrastructure and services throughout Africa. These initiatives focus on developing robust infrastructure, standardizing telecommunications services, and supporting emerging technologies that can propel Africa into the future.

7.2.1 Strategic Initiatives to Promote Digital Infrastructure and Services

7.2.1.1 Infrastructure Development

The ATU has been a key advocate for the expansion of digital infrastructure, including the development of fiber-optic networks, data centers, and internet exchange points (IXPs). By fostering partnerships between governments, private companies, and international organizations, the ATU has supported projects that enhance connectivity and improve the quality of digital services across the continent. These investments in infrastructure are critical to bridging the digital divide and ensuring that all Africans can benefit from the digital revolution.

7.2.1.2 Standardization and Interoperability

The ATU has strongly emphasized the importance of standardization and interoperability in telecommunications. By promoting the use of common technical standards, the Union aims to create a seamless digital environment where services and devices can function harmoniously across borders. This focus on standardization ensures that African consumers and businesses can enjoy a consistent and reliable digital experience, regardless of location.

7.2.1.3 Support for Emerging Technologies

The ATU has also been a strong advocate for the adoption of emerging technologies such as artificial intelligence (AI), blockchain, and the Internet of Things (IoT). These technologies have the potential to drive significant economic growth and innovation across various sectors. The ATU has facilitated discussions on the implementation of these technologies and encouraged member states to explore their applications for sustainable development. This support for emerging technologies includes [specific initiatives or programs], demonstrating the Union's forward-thinking approach and vision for Africa's digital future.

7.3 Collaboration with International Organizations and the Private Sector

Collaboration has been a cornerstone of the ATU's approach to driving digital innovation. The Union has formed strategic partnerships with international organizations such as the ITU, the African Union, and the World Bank, leveraging their expertise and resources to promote policies that support Africa's digital transformation. Through these partnerships, the ATU has been able to tackle the unique challenges facing the continent, from infrastructure development to Cybersecurity.

The ATU has also worked closely with the private sector to drive technological advancement and entrepreneurship. By engaging telecommunications companies, tech firms, and start-ups, the Union has facilitated the development of innovative solutions tailored to Africa's specific needs. These collaborations have resulted in investments in research and development and the creation of homegrown solutions that address Africa's challenges and harness its opportunities.

Furthermore, the ATU has hosted forums, workshops, and conferences to bring together stakeholders from various sectors. These platforms provide opportunities for networking, collaboration, and the exchange of ideas, ultimately contributing to the growth of Africa's digital economy.

The ATU's comprehensive approach to digital innovation underscores its commitment to shaping Africa's telecommunications future in a way that fosters economic growth, regional integration, and improved quality of life for all Africans.

CHAPTER EIGHT: KEY POLICIES AND FRAMEWORKS IMPLEMENTED BY THE AFRICAN TELECOMMUNICATIONS UNION (ATU)

8.1 Telecommunications Policy Harmonization

8.1.1 Development of Unified Telecommunications Policies Across Member States

The ATU has been a driving force in harmonizing telecommunications policies among its member states, understanding that a fragmented regulatory environment can hinder growth. By promoting the alignment of national telecommunications regulations with continental standards, the ATU is steadfast in its mission to forge a unified and integrated telecommunications landscape in Africa.

Through close collaboration with national governments, the ATU has developed a framework that addresses common regulatory issues, such as technological disparities, regulatory inconsistencies, and market fragmentation. By presenting a unified front to telecommunications stakeholders, the goal is to enable cross-border communication, foster economic cooperation, and attract investment.

A key moment in this process was the "African Telecommunications Policy Conference" held in 2018, a conference that was spearheaded by the ATU and brought together policymakers, industry leaders, and experts. This conference underscored the necessity for unified policies to drive regional cooperation, improve service delivery, and stimulate infrastructure investment. It was a significant step toward enhancing telecommunications integration across the continent.

8.1.2 Key Regulatory Frameworks Promoted by the ATU

In pursuit of its goal to harmonize policies, the ATU has introduced several regulatory frameworks essential to the effective management of Africa's telecommunications sector:

8.1.2.1 Spectrum Management

To ensure efficient use of radio frequencies, the ATU has provided member states with comprehensive guidelines for spectrum management. These guidelines facilitate coordinated frequency allocation, reduce interference, and support the deployment of emerging technologies like 5G and IoT. This management approach optimizes the use of limited resources while ensuring technological advancement.

8.1.2.2 Roaming Regulations

The ATU has taken significant steps to alleviate consumers' financial burdens through roaming regulations. By advocating for regional agreements, the Union is actively working to reduce roaming charges, making cross-border communication more affordable. This initiative is a testament to the ATU's commitment to strengthening regional connectivity and fostering economic ties among African nations.

8.1.2.3 Infrastructure Sharing

Infrastructure sharing has been another key focus for the ATU. By encouraging telecommunications operators to share infrastructure such as towers and fiber-optic cables, the Union aims to reduce operational costs and expand network coverage, particularly in underserved areas. This collaborative approach enhances the availability and affordability of telecommunications services across Africa.

8.2 Digital Infrastructure Development

8.2.1 ATU's Role in Promoting the Expansion of Broadband Infrastructure

The ATU has been instrumental in advocating for the expansion of broadband infrastructure across Africa, recognizing its role as a driver of economic development and a bridge to the digital economy. Through its "Broadband Infrastructure Development Plan," launched in partnership with the African Union, the ATU has promoted investment-friendly policies and public-private partnerships (PPPs) that encourage operators to expand broadband access.

By reducing regulatory barriers and fostering a conducive environment for investment, the ATU is accelerating the deployment of high-speed internet services, particularly in rural and underserved regions. The Union's efforts aim to connect millions of Africans to the digital world, fostering greater access to education, healthcare, and economic opportunities.

8.2.2 Efforts to Connect Underserved and Rural Areas

To address the digital divide, the ATU has prioritized initiatives that focus on connecting underserved and rural areas to modern telecommunications services:

8.2.2.1 Universal Service Funds (USFs)

The ATU has encouraged member states to establish Universal Service Funds (USFs) to support the development of telecommunications infrastructure in remote areas. These funds help subsidize network deployment, ensuring that even the most isolated communities can access basic communication services.

8.2.2.2 Community Networks

The Union has been a strong advocate for the creation of community networks—locally owned and operated services tailored to the specific needs of rural areas. These networks, often utilizing low-cost technologies and local resources, are a beacon of hope for communities that might otherwise be left behind, providing them with sustainable and affordable connectivity.

8.2.2.3 Public-Private Partnerships

Through fostering partnerships between governments, telecom operators, and international organizations, the ATU has facilitated joint investments in infrastructure that extend telecommunications services to rural areas. These partnerships not only focus on physical infrastructure but also emphasize capacity building and knowledge exchange to ensure the sustainability of these efforts.

8.3 Capacity Building and Education

8.3.1 Enhancing the Technical Expertise of Telecommunications Professionals in Africa

The ATU recognizes that the growth of the telecommunications sector in Africa is contingent upon a skilled workforce. In response, the Union has implemented a range of training programs that target key areas such as network management, cybersecurity, and regulatory practices. These initiatives are designed to equip telecommunications professionals with the expertise needed to navigate the sector's evolving landscape.

By partnering with regional training institutions and universities, the ATU ensures that these programs reflect the latest industry needs. The Union also organizes workshops and conferences that provide professionals with a platform to share knowledge, discuss emerging trends, and tackle the challenges posed by new technologies.

8.3.2 Partnerships with Educational Institutions and International Bodies for Training and Knowledge Exchange

In addition to its initiatives, the ATU has formed partnerships with leading educational institutions and international organizations to enhance training opportunities. Collaborations with bodies like the African Institute for Mathematical Sciences (AIMS) and the International Telecommunication Union (ITU) provide professionals with access to specialized training and globally recognized certification programs.

These partnerships support the establishment of regional training centers of excellence, which focus on nurturing talent and equipping Africa's telecommunications workforce with the skills necessary for digital transformation.

8.4 Collaborations with International Bodies

The ATU has recognized the importance of strategic alliances with international organizations to drive telecommunications development across the continent. Key partnerships with institutions such as the International Telecommunication Union (ITU), the World Bank, and the African Development Bank (AfDB) provide the technical expertise, financial resources, and innovative solutions needed to overcome Africa's unique challenges.

8.4.1 International Telecommunication Union (ITU)

The ATU collaborates closely with the ITU, leveraging the latter's global leadership in telecommunications policy and regulation. Together, they have organized workshops and initiatives focused on spectrum management and digital transformation, empowering African countries to harness telecommunications for socio-economic development.

8.4.2 World Bank

Through its partnership with the World Bank, the ATU has secured funding for critical telecommunications infrastructure projects, particularly in rural areas. The "Regional Communications Infrastructure Program" (RCIP) is one such initiative that has provided financial support to enhance broadband connectivity across Africa.

8.4.3 African Development Bank (AfDB)
The AfDB has been a key partner in promoting telecommunications infrastructure development through initiatives like the "Connect Africa" project. This collaboration focuses on increasing broadband penetration and building the capacity of local talent to ensure the sustainability of Africa's telecommunications networks.

8.5 Collaboration for Funding

8.5.1 Project Financing
Through joint efforts with international organizations like the World Bank and AfDB, the ATU has secured funding for projects aimed at expanding telecommunications infrastructure and improving service delivery. These initiatives not only provide financial resources but also promote regional cooperation among African countries.

8.5.2 Capacity Building Programs
Collaboration with international bodies has also enabled the ATU to offer training programs that build local capacity in telecommunications. These initiatives equip professionals with the skills needed to manage and operate Africa's growing telecommunications networks.

8.5.3 Engagement of Multinational Telecommunications Companies
Multinational telecommunications companies have played a pivotal role in Africa's telecommunications development, particularly through infrastructure

investment and technological innovation. Companies like MTN, Vodacom, and Airtel have made significant contributions to network expansion, service improvement, and the introduction of new technologies.

8.5.4 The Rise of Public-Private Partnerships (PPPs) in Telecommunications

Public-private partnerships (PPPs) have become a key model for improving connectivity in Africa. These partnerships combine the innovation of the private sector with the regulatory oversight of the public sector, facilitating the development of telecommunications infrastructure and improving service delivery.

By attracting investment and encouraging collaboration, PPPs have enabled governments to extend telecommunications services to underserved areas, ensuring that all Africans can benefit from modern communication technologies.

CHAPTER NINE: ONGOING CHALLENGES AND FUTURE OUTLOOK

The African telecommunications sector, despite notable progress in recent years, continues to grapple with challenges that hinder its growth and effectiveness. These challenges affect the delivery and accessibility of telecommunications services across the continent, particularly in rural and underserved areas.

9.1 Ongoing Infrastructure Gaps and the Digital Divide

One of the most pressing issues remains the infrastructure gap, particularly between urban and rural regions. While major cities enjoy advanced telecommunications networks, rural areas often lack basic services, exacerbating the digital divide. According to the International Telecommunication Union (ITU), in 2021, mobile penetration in many African cities exceeded 100%, while in rural areas, access rates fell below 50% in some countries.

1. Infrastructure Investment: Expanding telecommunications infrastructure in rural areas requires substantial investment. Building additional cell towers, extending fiber-optic networks, and ensuring a reliable power supply for telecommunications equipment are crucial

steps. However, the high costs of these projects deter private investors and strain government budgets.
2. Impact of the Digital Divide: This digital divide limits access to essential services, education, and economic opportunities. A report by the African Development Bank (AfDB) highlights how regions with limited internet access struggle to participate in the digital economy, thus stifling their development potential.

9.1.1 Regulatory and Policy Challenges in Adapting to New Technologies

As technology evolves rapidly, African nations face significant regulatory and policy challenges in keeping up with innovations such as 5G, Artificial Intelligence (AI), and the Internet of Things (IoT). These challenges may prevent the continent from fully benefiting from these advancements.

1. Regulatory Frameworks: Many African countries lack robust regulatory structures capable of overseeing new telecommunications technologies. The rapid pace of innovation often outstrips regulators' ability to develop effective policies. For instance, implementing 5G requires new regulations concerning spectrum allocation and network sharing, areas where many countries are still lagging.
2. Investment and Adoption: Uncertain regulatory environments also deter investors from committing to these new technologies. Telecommunications operators often face bureaucratic delays in obtaining necessary

licenses and permits, slowing the adoption of innovative services.

9.2 Financial Constraints and Reliance on Foreign Aid

Many African nations' financial limitations continue to hinder the development of their telecommunications infrastructure, leading to a heavy dependence on foreign aid and investment.

1. Funding Gaps: The funding required to close the telecommunications infrastructure gap in Africa is estimated to be in the billions. According to ITU reports, achieving universal broadband access will need an estimated $100 billion in annual investment. However, budgetary constraints in many African nations prevent them from meeting this substantial demand.
2. Dependence on Foreign Aid: The reliance on foreign aid introduces complications, as donor-driven projects may not always align with national priorities. Additionally, excessive reliance on foreign aid can stifle local innovation and limit long-term sustainability in the sector.

9.3 The Future of the ATU

The African Telecommunications Union (ATU) is positioned to play a critical role in shaping the future of Africa's telecommunications sector. As the landscape evolves, the ATU must address both emerging challenges and opportunities to ensure that telecommunications development benefits all Africans.

1. Strategic Goals for the Future: The ATU has outlined strategic goals to expand digital infrastructure, promote innovation, and achieve universal connectivity across the continent.
2. Expanding Digital Infrastructure: One of the ATU's core goals is to facilitate the expansion of digital infrastructure, especially in underserved areas. This includes promoting investments in broadband networks, enhancing mobile connectivity, and ensuring reliable power sources for telecommunications infrastructure.
3. Promoting Innovation: The ATU aims to foster innovation in the telecommunications sector by encouraging research and development, as well as supporting tech startups. By creating a conducive environment for innovation, the ATU seeks to position African nations as key players in the global digital economy.
4. Achieving Universal Connectivity: The ATU is committed to achieving universal connectivity, advocating for policies that ensure equitable access to telecommunications services. This involves promoting affordable pricing, reducing regulatory barriers, and encouraging public-private partnerships to address infrastructure gaps.

9.4 The ATU's Vision for Africa's Telecommunications Sector in the Next Decade

Looking ahead, the ATU envisions a telecommunications sector that is robust, innovative, inclusive, and equitable. Several key priorities will guide this vision:

1. Digital Inclusion: The ATU will continue to champion digital inclusion, ensuring that marginalized communities have access to telecommunications services. This includes directing investments to rural areas, promoting digital literacy programs, and advocating for policies that support underserved populations.
2. Cybersecurity: As Africa's digital landscape expands, so do cybersecurity risks. The ATU is focused on developing strategies and frameworks to safeguard telecommunications networks and protect users from cyber threats. Collaboration with international organizations will be essential to sharing best practices and enhancing network security.
3. Technological Sovereignty: The ATU seeks to ensure that African nations can develop and control their telecommunications technologies. This involves investing in local research and development, building partnerships with technology providers, and fostering the growth of homegrown solutions.

9.5 How the ATU Aims to Address Emerging Challenges

The ATU recognizes the importance of addressing emerging challenges to ensure the sustainable growth of Africa's telecommunications sector. Several initiatives are underway to tackle these issues.

1. Digital Inclusion Initiatives: The ATU plans to launch targeted initiatives that increase access to telecommunications services in underserved communities. This may involve working with member states to develop policies that incentivize private sector investment in rural areas and facilitate access to affordable devices and services.
2. Cybersecurity Frameworks: To address growing cybersecurity threats, the ATU will establish comprehensive frameworks that safeguard telecommunications networks. These efforts include developing guidelines for network operators, promoting cybersecurity awareness, and fostering collaboration between member states to share information on cyber threats and vulnerabilities.
3. Capacity Building for Technological Sovereignty: The ATU aims to strengthen Africa's technical expertise by offering training and resources to local professionals. This will help build a skilled workforce capable of developing and managing telecommunications technologies, reducing reliance on foreign expertise and enhancing Africa's technological autonomy.

In summary, while challenges remain, the future of Africa's telecommunications sector is bright, driven by the ATU's commitment to expanding digital infrastructure, fostering innovation, and ensuring that all Africans benefit from the digital revolution. The path forward involves not only overcoming the current hurdles but also embracing emerging opportunities for sustainable growth and development.

CHAPTER TEN: THE BROADER IMPACT OF THE ATU ON AFRICAN SOCIETY

The African Telecommunication Union (ATU) has been a key player in shaping African society. Its establishment and ongoing efforts have led to the emergence of telecommunications as a transformative force, reshaping Education, healthcare, and governance. This has driven social and economic development across the continent, with the expansion of digital services opening up new potential for growth, inclusion, and empowerment.

10.1: The Transformative Impact of Telecommunications on Education, Healthcare, and Governance

10.1.1 Education

The integration of telecommunications into education systems has been a game-changer for learning across Africa. E-learning platforms and mobile applications have broadened access to Education, especially for students in remote and underserved regions. The African Virtual University (AVU), for instance, has harnessed telecommunications to provide distance learning across multiple countries, making quality education accessible beyond physical classrooms. According to a report by the International Telecommunication Union (ITU), mobile learning initiatives in countries like Kenya have boosted rural school enrollment rates by 30%. These efforts have been crucial in bridging educational gaps and equipping the next generation with the skills needed to thrive in the digital age, inspiring a wave of change in the African education landscape.

10.1.2 Healthcare

Telecommunications have also revolutionized healthcare delivery, particularly in regions with limited medical infrastructure. Telemedicine services have grown exponentially, providing critical healthcare access to rural populations. In Uganda, for instance, mobile phones now link patients with healthcare providers, reducing patient wait times for consultations by 25%. Additionally, mobile health (mHealth) initiatives have proved invaluable during public health emergencies, such as the COVID-19 pandemic, when rapid communication of health information saved lives. These innovations have been instrumental in improving health outcomes and ensuring that even remote communities can benefit from modern medical services, offering a reassuring glimpse into the future of healthcare in Africa.

10.1.3 Governance

Telecommunications are transforming governance by promoting transparency, accountability, and citizen engagement. E-governance platforms, such as Rwanda's "e-Government" initiative, allow citizens to access government services online, reducing bureaucratic inefficiencies and improving service delivery. These platforms foster trust between governments and citizens by increasing transparency and facilitating direct communication. As African governments continue to embrace digital tools, the role of telecommunications in governance will only grow, further empowering citizens to participate in public decision-making and fostering an optimistic outlook on the future of governance in Africa.

10.1.4 Case Studies of Telecommunications' Impact on Quality of Life

10.1.4.1 Kenya's M-Pesa

The introduction of M-Pesa, a mobile money transfer service, has revolutionized financial inclusion in Kenya. Since its launch in 2007, M-Pesa has given millions of previously unbanked individuals—especially in rural areas—the ability to conduct financial transactions. As of 2021, M-Pesa had over 40 million active users, contributing to increased household incomes and economic resilience across the country. This mobile platform has not only redefined banking in Kenya but also inspired similar innovations across the continent.

10.1.4.2 South Africa's Education Initiatives

In South Africa, telecommunications have transformed Education through the "Smart School" initiative, which provides schools with internet access and digital learning tools. This program has significantly enhanced student engagement and academic performance. Within two years of its implementation, participating schools reported a 15% increase in exam scores, highlighting the profound impact of digital tools on educational outcomes.

10.1.4.3 Telemedicine in Nigeria

In Nigeria, telemedicine platforms are addressing healthcare shortages in rural areas. The "Rural Telemedicine Project" uses mobile technology to connect healthcare professionals with patients in remote communities, significantly improving healthcare delivery. This initiative has led to higher vaccination rates and better management of chronic diseases, showcasing the power of telecommunications in enhancing public health.

10.2 Empowerment Through Connectivity

The ATU's initiatives have significantly contributed to expanding telecommunications access, particularly for marginalized communities. By bridging the digital divide, the ATU is driving socio-economic growth and unlocking the potential of millions across the continent.

10.2.1 Expanding Access

One of the ATU's key achievements has been increasing access to telecommunications services for marginalized populations, including women, youth, and rural communities. Through partnerships with governments and NGOs, the ATU has worked to promote infrastructure development in underserved areas. Notably, initiatives aimed at increasing women's access to mobile technology have seen a surge in female entrepreneurs using telecommunications to grow their businesses, driving both personal and economic development.

10.2.2 Digital Literacy Programs

The ATU has been instrumental in promoting digital literacy, equipping communities with the skills necessary to navigate the digital landscape. These programs have been particularly effective in empowering women and youth, who are often left behind in technological advancements. A World Bank report noted that digital literacy programs have boosted employment opportunities by 40% in participating communities, further underscoring the transformative power of telecommunications in building sustainable livelihoods.

10.2.3 Driving Youth Entrepreneurship, Innovation, and Social Mobility

10.2.3.1 Youth Entrepreneurship

Telecommunications have become a powerful engine for youth entrepreneurship in Africa. Young innovators are leveraging mobile technology to launch businesses, connect with markets, and access customers. Platforms like Jumia and Konga have opened doors for e-commerce, allowing youth-led enterprises to flourish and contribute to economic growth. These ventures are driving job creation and offering new pathways for social mobility.

10.2.3.2 Innovation Hubs

The ATU has supported the establishment of innovation hubs across Africa, where young entrepreneurs can develop and implement tech-driven solutions. The "African Innovation Hub" initiative has led to the creation of more than 100 tech incubators that foster innovation and knowledge sharing. These hubs are breeding grounds for creativity, propelling Africa's tech scene forward and positioning the continent as a global player in digital innovation.

10.2.3.3 Social Mobility

Access to telecommunications has been closely linked with improved social mobility. Research indicates that communities with better telecommunications infrastructure experience higher levels of economic activity, enabling individuals to break free from poverty and improve their living conditions. By promoting inclusive policies, the ATU is ensuring that marginalized populations can participate in the digital economy and improve their socio-economic status.

In summary, the African Telecommunication Union has played a pivotal role in shaping Africa's socio-economic landscape. Through the promotion of telecommunications in Education, healthcare, and governance, the ATU has improved the quality of life for millions of people. Its efforts to expand access, empower marginalized communities, and foster innovation have not only driven economic growth but also paved the way for a more connected, inclusive, and equitable future for Africa. As the continent continues to embrace digital technologies, the ATU will remain a cornerstone of Africa's ongoing development, ensuring that all Africans benefit from the digital revolution.

CHAPTER ELEVEN: THE BROADER IMPACT OF THE ATU ON AFRICAN SOCIETY

The African Telecommunication Union (ATU) has played a transformative role in shaping African society. Since its inception, the ATU's initiatives, such as the expansion of telecommunications infrastructure, the promotion of e-learning platforms, and the implementation of telemedicine services, have not only expanded telecommunications infrastructure but have also driven socio-economic development across the continent. By integrating telecommunications into critical sectors like Education, healthcare, and governance, the ATU has empowered marginalized populations, enhanced access to essential services, and laid the foundation for sustainable growth.

11.1 The Transformative Impact of Telecommunications on Human Development and Governance

10.1.1 Education

Telecommunications have revolutionized Education in Africa, turning what was once a privilege for the few into a broader right for many. E-learning platforms and mobile applications have provided new avenues for learning, especially in remote and underserved regions. The African Virtual University (AVU), a pioneer in distance education, has harnessed the power of telecommunications to break down barriers that once limited access to quality education. This effort has opened doors for students across Africa, connecting them to resources and opportunities that would have otherwise been out of reach. A study by the International Telecommunication Union (ITU) reveals that in countries like Kenya, mobile learning initiatives have increased rural school enrollment rates by 30%. This surge is a testament to the role of telecommunications in creating an educated, skilled generation capable of driving Africa's future development.

11.1.2 Healthcare

Telecommunications have also reshaped healthcare, making medical services more accessible to underserved populations. In regions where healthcare infrastructure is sparse, telemedicine has emerged as a lifeline. Uganda's telemedicine services, for example, have reduced patient wait times by 25%, allowing quicker access to medical consultations via mobile technology. Mobile health (mHealth) initiatives have been crucial in disseminating public health information during emergencies such as the COVID-19 pandemic. By using mobile platforms, governments were able to rapidly communicate health protocols, distribute updates, and deliver life-saving advice to millions. As telecommunications continue to penetrate deeper into rural areas, the outlook for healthcare delivery in Africa becomes more promising.

11.1.3 Governance

Telecommunications have introduced a new era of governance in Africa, characterized by transparency, accountability, and increased citizen engagement. E-governance platforms, such as Rwanda's "e-Government" initiative, have enabled citizens to access government services online, reducing the inefficiencies of traditional bureaucracy. These platforms not only streamline service delivery but also foster trust between governments and citizens by allowing direct communication and ensuring that the government is more responsive to the needs of the public. As more African nations adopt digital governance tools, telecommunications will continue to drive democratic participation and strengthen the relationship between governments and their citizens.

11.1.4 Case Studies of Telecommunications' Impact on Quality of Life

Telecommunications' ability to improve the quality of life in Africa is a beacon of hope that can be observed through various case studies:

11.1.4.1 Kenya's M-Pesa

M-Pesa, Kenya's mobile money service, has transformed financial inclusion for millions. Since its inception in 2007, it has allowed those without access to traditional banking to perform financial transactions via their mobile phones. As of 2021, M-Pesa had more than 40 million active users, providing economic resilience to rural communities and enabling them to increase household incomes. Its success has inspired similar mobile banking models across Africa, helping millions engage in the formal economy for the first time.

11.1.4.2 South Africa's Education Initiatives

South Africa's 'Smart School' initiative, for instance, equips schools with internet access and digital resources, thereby enhancing educational outcomes. This initiative has driven significant improvements in student performance, with participating schools reporting a 15% increase in exam scores within just two years. The initiative also provides teachers with training on how to effectively use digital resources in their teaching, further enhancing the quality of Education. This case highlights how digital tools can bridge gaps in educational quality and student engagement across the continent.

11.1.4.3 Telemedicine in Nigeria

In Nigeria, the "Rural Telemedicine Project" connects healthcare providers with patients in remote areas via mobile technology. This initiative has improved access to healthcare services and led to positive health outcomes, including increased vaccination rates and improved management of chronic diseases. Telecommunications have thus become an essential tool for addressing healthcare disparities and enhancing the quality of life for those in underserved communities.

11.2 Empowerment Through Connectivity

The ATU's commitment to expanding telecommunications has not only empowered millions, particularly those in marginalized communities but also served as a beacon of progress and potential. By bridging the digital divide, the ATU has fostered socio-economic growth, opened new avenues for innovation, and created opportunities for inclusion.

11.2.1 Expanding Access

A cornerstone of the ATU's mission is ensuring that telecommunications services are accessible to all, especially women, youth, and rural communities. Through partnerships with local governments and NGOs, the ATU has spearheaded infrastructure projects aimed at bringing connectivity to underserved areas. Programs that focus on women's access to mobile technology have yielded significant results, with a notable rise in female entrepreneurs who have leveraged telecommunications to grow their businesses. This shift has driven personal and economic development, contributing to the financial independence of women across the continent.

11.2.2 Digital Literacy Programs

The ATU has also championed digital literacy programs to equip people with the skills needed to navigate the digital world. These programs, which include training on how to use digital tools for business, communication, and Education, have been particularly impactful for youth and women, who have traditionally been excluded from technological advancements. According to a World Bank report, communities that participated in digital literacy programs saw a 40% increase in employment opportunities. The ATU's focus on digital literacy has not only empowered individuals by providing them with the skills they need to succeed in the digital age but also strengthened communities by creating a more tech-savvy, economically resilient population.

11.2.3 Driving Youth Entrepreneurship, Innovation, and Social Mobility

Telecommunications have sparked a new wave of youth entrepreneurship, innovation, and social mobility across Africa.

11.2.3.1 Youth Entrepreneurship

Young innovators are harnessing telecommunications to launch businesses and connect with global markets. Platforms like Jumia and Konga have opened new frontiers for e-commerce, enabling youth-led companies to thrive. This entrepreneurial drive is generating jobs and creating economic opportunities for a new generation of African leaders.

11.2.3.2 Innovation Hubs

The ATU's support for innovation hubs across Africa has catalyzed the creation of more than 100 tech incubators where young entrepreneurs can develop tech-driven solutions. These hubs are hotbeds of creativity and play a crucial role in positioning Africa as a growing force in the global digital economy.

11.2.3.3 Social Mobility

Improved telecommunications access has also led to increased social mobility. Studies show that communities with better telecommunications infrastructure experience heightened economic activity, helping people escape poverty and improve their living conditions. The ATU's efforts to promote inclusive policies ensure that even the most marginalized groups can benefit from these advancements and participate in the digital economy.

In summary, the African Telecommunication Union has had a profound impact on Africa's social and economic development. Through its dedication to expanding telecommunications, the ATU has enhanced Education, healthcare, and governance, improving the quality of life for millions. Its efforts to empower marginalized communities and foster digital inclusion have unleashed new waves of innovation and entrepreneurship, paving the way for a more connected, inclusive, and equitable future for Africa. As the continent continues to embrace digital technologies, the ATU will remain a vital force in ensuring that all Africans can thrive in the digital age.

CHAPTER TWELVE: STRUCTURE AND GOVERNANCE OF THE PROPOSED PAN-AFRICAN TELECOMMUNITY (PAT)

The success of the proposed Pan-African Telecommunity (PAT) will depend heavily on its organizational structure, decision-making processes, and mechanisms for ensuring fair representation. This chapter presents a comprehensive framework for the PAT, drawing from successful regional and international telecommunications organizations. It is important to note that this framework has been carefully adapted to suit Africa's unique needs and context, taking into account the region's diverse telecommunications landscape and the challenges it faces.

12.1 Organizational Framework

The PAT's structure must be designed to promote efficiency, foster collaboration, and ensure that all member states have a voice. The proposed framework includes several key components:

12.1.1 General Assembly

The General Assembly will serve as the highest decision-making body within the PAT, providing strategic direction and approving major policies. Its structure and responsibilities will include:

1. Composition: All member states will be represented.
2. Frequency: Annual meetings to discuss strategic initiatives and vote on important issues.
3. Duties: Electing the Executive Council and Secretary-General, among other key roles.

This model is inspired by the International Telecommunication Union (ITU), whose global governance system has been effective in driving telecommunications policy.

12.1.2 Executive Council

The Executive Council will oversee policy implementation and ensure that the PAT's strategies are executed efficiently. It will:

1. Composition: 15-20 representatives from member states, elected on a rotating basis to maintain regional balance.
2. Meetings: Held quarterly or as necessary.
3. Focus: Responsible for reviewing progress on initiatives and providing oversight.

The Executive Council will operate similarly to that of the African Union, ensuring regional representation and collaboration across Africa's diverse telecommunications landscape.

12.1.3 Secretariat

The Secretariat will serve as the engine driving the PAT's day-to-day operations. Key aspects include:

1. Leadership: Headed by the Secretary-General, elected for a four-year term.

2. Functions: Responsible for implementing the decisions of the General Assembly and Executive Council, managing the organization, and overseeing departments such as Technical Standards, Spectrum Management, and Digital Development.

The Secretariat will be modeled after the GSMA, which has been effective in managing mobile operator associations globally.

12.1.4 Specialized Committees

Specialized Committees will be formed to focus on critical areas such as spectrum management, cybersecurity, and digital inclusion. These committees will:

1. Composition: Experts from member states and the telecommunications industry.
2. Duties: Provide expert recommendations to the Executive Council.
3. Report: Regularly to the Executive Council on their areas of focus.

The European Conference of Postal and Telecommunications Administrations (CEPT) offers a successful precedent for a similar committee structure.

12.1.5 Regional Offices

To ensure that policies are effectively implemented across the continent, regional offices will be established in each of Africa's five regions (North, South, East, West, and Central). These offices will:

1. Roles: Serve as regional hubs to implement PAT initiatives and coordinate with national regulatory authorities.
2. Function: Facilitate communication between regional governments and the PAT.

The International Telecommunication Union's regional office model will serve as a guide for this structure.

12.2 Decision-Making Processes

For the PAT to function efficiently, it must adopt a decision-making process that is inclusive, transparent, and agile. The following methods are proposed, ensuring that all member states are involved and their voices are heard:

12.2.1 Consensus-Based Approach

The PAT will aim to achieve consensus on all major policy decisions. If consensus is unattainable:

1. Voting Process: Voting will be used as a fallback method.
2. Advantages: A consensus-based approach promotes buy-in from all member states.

This system is inspired by the World Trade Organization's model, which emphasizes consensus in its decision-making processes.

12.2.2 Tiered Decision-Making

The PAT's decision-making framework will be tiered to ensure efficiency:

1. Routine Decisions: Managed by the Secretariat.
2. Policy Decisions: Overseen by the Executive Council.

3. Strategic Decisions: These are decided by the General Assembly, including elections and major policy shifts.

This tiered structure ensures that decisions are made at the appropriate level, preventing bureaucratic gridlock.

12.2.3 Transparent Deliberation

Transparency is critical for the credibility of the PAT. Proposed transparency measures include:

1. Open Meetings: General sessions will be open to the public, with provisions for closed sessions on sensitive matters.
2. Publication of Decisions: Minutes and policy documents will be made publicly available.
3. Public Consultations: Major policy changes will undergo public consultation before implementation.

The European Telecommunications Standards Institute (ETSI) provides a model for transparency that the PAT can adapt.

12.2.4 Appeal Mechanism

To maintain accountability, the PAT will have an independent appeal body. This body will:

1. Dispute Resolution: Handle disputes between member states and the PAT's governing bodies.
2. Process: Establish clear procedures for appeals and decision challenges.

The World Bank's Inspection Panel serves as a useful model for such an appeal mechanism.

12.3 Representation and Voting Rights

Ensuring fair representation and voting rights is essential for the PAT's long-term success. A balanced system will guarantee that all member states, regardless of size or economic power, have a voice.

12.3.1 Universal Membership

The PAT will have a system of universal membership where:

1. Eligibility: All African states are eligible to join as full members.
2. Associate Membership: Industry stakeholders and international organizations can hold non-voting associate memberships, providing valuable input.

This structure mirrors the United Nations model of inclusive membership.

12.3.2 One Country, One Vote

To ensure equality:

1. Voting: Each member state will have one vote in the General Assembly.
2. Equity: This system guarantees equal representation, regardless of a country's size or telecommunications market.

This principle was successfully used in the United Nations General Assembly.

12.3.3 Weighted Voting for Certain Decisions

For decisions with significant economic implications, such as those related to financial investments or market regulations:

1. Weighted Voting: Voting could be weighted based on population, GDP, or telecommunications market size.
2. Periodic Reviews: The weight of each vote would be reviewed periodically to reflect changing economic realities.

The International Monetary Fund's weighted voting system provides a model for how this could work.

12.3.4 Regional Balance

To ensure fair representation:

1. Regional Representation: The Executive Council and key committees must include representatives from each of Africa's five regions.
2. Leadership Rotation: Leadership positions will rotate among the regions.

This principle, modeled after the African Union, ensures that no region dominates decision-making.

12.3.5 Industry Representation

Industry stakeholders will play a key role in the PAT's governance:

1. Advisory Boards: Advisory boards will be formed with representatives from telecommunications companies, equipment manufacturers, and civil society organizations.

2. Consultative Role: These boards will provide input without voting rights, similar to the multi-stakeholder model used by ICANN.

12.3.6 Quorum and Majority Requirements

To prevent deadlock and ensure decisions can be made efficiently:

1. Quorum: A two-thirds quorum will be required for the General Assembly.
2. Majority Rules: A simple majority will suffice for routine decisions, while major policy shifts will require a two-thirds majority.

In summary, the proposed structure and governance of the Pan-African Telecommunity (PAT) are designed to balance efficiency, inclusivity, and transparency. Drawing on best practices from successful international organizations, the PAT's framework will empower Africa's telecommunications sector and drive sustainable development. As the PAT evolves, its structure must remain flexible, adapting to the rapidly changing technological landscape to stay relevant and effective in addressing the needs of African nations. Regular reviews and reforms will be key to ensuring that the PAT continues to serve as a driving force for Africa's digital transformation.

CHAPTER THIRTEEN: KEY OBJECTIVES AND RESPONSIBILITIES OF THE PAN-AFRICAN TELL COMMUNITY (PAT)

The Pan-African Telecommunity (PAT) is committed to shaping the future of telecommunications across Africa. This chapter outlines the essential objectives and responsibilities that the PAT should pursue to establish a robust, secure, and inclusive digital ecosystem for the continent. By focusing on policy standardization, spectrum management, and cybersecurity, the PAT can ensure Africa's telecommunications sector not only survives but thrives, adapting to and even leading new technological challenges.

13.1 Standardization of Telecommunications Policies

A key objective of the PAT will be to harmonize telecommunications policies across Africa, creating a cohesive and efficient regulatory environment. This approach will foster cross-border services, encourage investments, and reduce fragmentation that hinders the industry's growth.

13.1.1 Regulatory Harmonization:
1. Develop a unified regulatory framework for African member states.

2. Prioritize areas such as licensing, interconnection agreements, and quality-of-service standards.
3. Streamline regulatory processes to remove barriers to cross-border services and investments.

The European Union's (EU) regulatory framework for electronic communications serves as a model, demonstrating how harmonization can drive regional connectivity.

13.1.2 Digital Market Integration:
1. Facilitate the development of a single digital market for Africa, allowing seamless cross-border operations.
2. Encourage free and secure cross-border data flows while respecting each country's sovereignty.
3. Align efforts with the African Continental Free Trade Area (AfCFTA) to strengthen regional digital economies.

The EU's Digital Single Market strategy offers valuable insights into fostering continental digital integration.

13.1.3 Universal Service Policies:
1. Develop comprehensive guidelines for universal service policies aimed at expanding connectivity to underserved areas.
2. Advocate for innovative funding mechanisms, such as public-private partnerships, to support these efforts.

The International Telecommunication Union (ITU) provides useful guidelines for universal service funds, offering a global perspective.

13.1.4 Emerging Technology Regulation:
1. Craft forward-looking policies to regulate and facilitate the adoption of emerging technologies, including 5G, the Internet of Things (IoT), and artificial intelligence (AI).
2. Ensure these regulations are flexible and able to evolve with technological advancements.

The Organization for Economic Cooperation and Development (OECD) has valuable experience in developing adaptive policies for emerging technologies.

13.2 Spectrum Management and Allocation

Efficient spectrum management is essential to the growth of wireless communications across Africa. The PAT should coordinate spectrum policies and maximize the use of available frequencies to accelerate the digital revolution.

13.2.1 Regional Spectrum Planning:
1. Create a comprehensive spectrum plan for Africa that aligns with international standards and ensures efficient use.
2. Work with international bodies such as the ITU to guarantee global alignment and cooperation.
3. Facilitate cross-border coordination to reduce interference and optimize spectrum use.

The European Radio Spectrum Policy Programme serves as a model for regional spectrum coordination and planning.

13.2.2 Harmonized Band Plans:
1. Identify key frequency bands for services like mobile broadband and IoT, promoting harmonized use across Africa.
2. Enable economies of scale in equipment manufacturing and foster cross-border service provision.

The GSMA's band plan recommendations provide industry expertise on the benefits of harmonization.

13.2.3 Spectrum Sharing Frameworks:
1. Develop policies that encourage efficient spectrum usage through innovative sharing mechanisms.
2. Explore dynamic spectrum access and the use of unlicensed spectrum to maximize value.

Case studies from the U.S. Federal Communications Commission (FCC) demonstrate the potential of spectrum-sharing strategies.

13.2.4 Digital Dividend Strategies:
1. Coordinate the reallocation of spectrum freed up by the digital transition to maximize socio-economic benefits.
2. Use these digital dividends to enhance services like mobile broadband.

The ITU's digital dividend guidelines provide a foundation for crafting effective strategies.

13.3 Cybersecurity and Data Protection

As Africa's digital ecosystem grows, the PAT is dedicated to playing a critical role in establishing strong cybersecurity measures and protecting data. Coordinating continent-wide efforts will be vital for building trust and resilience in Africa's digital infrastructure, ensuring that stakeholders can operate with confidence in a secure environment.

13.3.1 Pan-African Cybersecurity Strategy:
1. Develop a comprehensive cybersecurity strategy tailored to Africa's needs.
2. Encourage the establishment of national Computer Emergency Response Teams (CERTs) across member states to address cyber threats.
3. Promote sharing of cybersecurity information to enhance collective security.

The African Union Convention on Cyber Security and Personal Data Protection provides a robust starting point for these efforts.

13.3.2 Critical Infrastructure Protection:
1. Establish guidelines to safeguard critical telecommunications infrastructure from physical and cyber threats.
2. Promote network designs that are resilient and incorporate redundancy to minimize service disruptions.

The EU's NIS Directive offers a strong model for protecting critical telecommunications infrastructure.

13.3.3 Data Protection and Privacy Framework
1. Develop a pan-African data protection framework that addresses both innovation and privacy concerns.
2. Align this framework with global standards, such as the EU's General Data Protection Regulation (GDPR), while also considering Africa's unique context.

The African Union's Personal Data Protection Guidelines provide a useful starting point for data protection policies.

13.3.4 Capacity Building in Cybersecurity:
1. Coordinate initiatives to build cybersecurity capacity and expertise throughout Africa.
2. Foster partnerships with global cybersecurity organizations and universities to offer training and develop local expertise.

The ITU's Global Cybersecurity Agenda offers tools and resources for capacity building in this area.

13.3.5 Child Online Protection:
1. Create and implement guidelines to protect children from the risks of the digital environment.
2. Promote digital literacy programs focused on educating children and families about online safety.

UNICEF's work on child online protection provides valuable insights into best practices for safeguarding young users.

In summary, the objectives and responsibilities outlined for the Pan-African Telecommunity (PAT) are ambitious but essential to realizing the full potential of Africa's telecommunications future. By focusing on policy harmonization, spectrum management, cybersecurity, and data protection, the PAT can lay the groundwork for a thriving, secure, and inclusive digital ecosystem across Africa. Flexibility will be critical as the technological landscape evolves, requiring regular review and updating of policies to meet the continent's ever-changing needs.

CHAPTER FOURTEEN: INFRASTRUCTURE DEVELOPMENT AND SHARING BY THE PROPOSED PAN-AFRICAN TELL COMMUNITY (PAT)

Establishing a robust telecommunications infrastructure is essential and a critical step towards achieving Africa's digital transformation. This chapter delves into strategies for broadening broadband access, enhancing cross-border connectivity, and leveraging shared satellite and undersea cable projects across the continent, all of which are key to this transformation.

14.1 Strategies for Expanding Broadband Access

Expanding broadband access is imperative for bridging the digital divide and catalyzing economic growth throughout Africa. The African Telecommunity Union (ATU), as a key player in the region's telecommunications sector, should take the lead in implementing the following strategies:

14.1.1 National Broadband Plans

1. Encouragement for Development: Inspire member states to craft and execute comprehensive national broadband plans that consider local needs and conditions.

2. Guidelines and Best Practices: Provide detailed guidelines and best practices for effective plan formulation and implementation, ensuring a systematic approach to broadband expansion.

The ITU's "Connect 2030 Agenda" serves as a valuable framework for national broadband planning, fostering collaboration and alignment among nations.

14.1.2 Infrastructure Sharing

1. Promote Sharing Policies: Advocate for policies that incentivize infrastructure sharing among telecommunications operators, thus minimizing redundancy and maximizing resource utilization.
2. Fair Agreement Development: Develop comprehensive guidelines to ensure that infrastructure-sharing agreements are fair, transparent, and beneficial for all parties involved.

The GSMA's research on infrastructure sharing in emerging markets provides crucial insights and best practices that can be adapted.

14.1.3 Public-Private Partnerships (PPPs)

1. Facilitating Partnerships: Facilitate the formation of PPPs to enhance infrastructure development, particularly in underserved regions where private investment is often limited.
2. Model Frameworks: Create adaptable model PPP frameworks that member states can tailor to their specific contexts and requirements.

The World Bank's PPP Knowledge Lab offers invaluable resources and best practices to guide successful partnerships.

14.1.4 Universal Service Funds
1. Optimizing Fund Usage: Maximize the effectiveness of Universal Service Funds (USFs) to bolster broadband expansion efforts in underserved areas.
2. Guidelines for Management: Establish comprehensive guidelines for effective fund management and project selection to ensure transparency and accountability.

The ITU's study on Universal Service Funds in Africa provides relevant case studies that can inform best practices.

14.1.5 Innovative Technologies
1. Promoting Adoption: Encourage the adoption of innovative technologies to improve last-mile connectivity, making internet access more affordable and widespread.
2. Exploring Solutions: Investigate and implement alternative solutions such as TV White Spaces, balloon-powered internet, and low-earth orbit satellites to reach remote areas.

Microsoft's Airband Initiative illustrates the potential of TV White Spaces technology in expanding broadband access.

14.2 Cross-Border Connectivity Initiatives

Strengthening cross-border connectivity is vital for realizing a digitally integrated Africa. The ATU should concentrate on the following initiatives:

14.2.1 Regional Backbone Networks
1. Coordinating Development: Lead the coordination of developing regional fiber optic backbone networks to facilitate seamless connectivity across borders.
2. Facilitating Agreements: Promote collaborative agreements for cross-border fiber deployment and management to enhance service delivery.

The World Bank's Digital Economy for Africa initiative provides key insights into successful regional connectivity projects.

14.2.2 One Network Area Initiatives
1. Expanding Initiatives: Broaden the scope of "One Network Area" initiatives to additional regions, effectively reducing roaming charges for users and enhancing mobility.
2. Mobile Money Frameworks: Establish frameworks for cross-border mobile money transactions, simplifying processes and boosting economic activities.

The East African Community's One Network Area exemplifies a successful model that can be replicated in other regions.

14.2.3 Internet Exchange Points (IXPs)
1. Establishing IXPs: Advocate for the creation of national and regional Internet Exchange Points to improve Internet routing efficiency and reduce costs.
2. Encouraging Peering Policies: Develop supportive policies that promote local and regional peering, fostering a more resilient internet ecosystem.

The Internet Society's initiatives on IXPs in Africa offer a wealth of resources to guide implementation.

14.2.4 Harmonized Rights of Way
1. Developing Policies: Formulate harmonized policies governing rights of way across borders, ensuring streamlined processes for deploying cross-border infrastructure.
2. Streamlining Processes: Work to simplify and standardize procedures to obtain rights of way and facilitate smoother project execution.

The African Development Bank's Program for Infrastructure Development in Africa (PIDA) addresses crucial cross-border infrastructure challenges.

14.3 Shared Satellite and Undersea Cable Projects

Collaborative initiatives for satellite and undersea cable projects can significantly boost Africa's international connectivity:

14.3.1 Pan-African Satellite Initiatives
1. Feasibility Exploration: Assess the feasibility of establishing a pan-African satellite system

aimed at enhancing broadband access and broadcasting capabilities across the continent.
2. Coordinated Spectrum Allocation: Synchronize spectrum allocation and orbital slot applications at the international level to ensure operational efficiency.

The European Space Agency's ARTES program provides insights into collaborative satellite initiatives that can inform Africa's approach.

14.3.2 Regional Satellite Projects
1. Facilitating Regional Projects: Promote the establishment of regional satellite projects tailored to meet the specific needs of geographic areas across Africa.
2. Equitable Cost Sharing: Develop frameworks that ensure equitable sharing of costs and benefits among participating nations, fostering cooperation and mutual gain.

The Arab Satellite Communications Organization (Arabsat) serves as a model for regional satellite collaboration.

14.3.3 Coordinated Undersea Cable Planning
1. Comprehensive Planning: Create a thorough plan for deploying undersea cables around Africa, focusing on maximizing connectivity and accessibility.
2. Landing Points Coordination: Coordinate landing points and terrestrial backhaul to ensure maximum impact and efficiency of cable networks.

The 2Africa undersea cable project exemplifies the potential benefits of large-scale collaborative efforts in enhancing connectivity.

14.3.4 Open Access Models
1. Promoting Open Access: Advocate for open access models for undersea cables and landing stations to ensure fair competition and broad access to international bandwidth.
2. Guideline Development: Establish guidelines that ensure non-discriminatory access to these essential resources, fostering innovation and growth.

The World Bank's study on open-access models in Africa offers valuable insights into best practices.

14.3.5 Capacity Building for Cable Management
1. Building Local Capacity: Implement programs to develop local expertise in the operation and maintenance of undersea cables, enhancing sustainability.
2. Knowledge Sharing: Encourage knowledge sharing among African countries regarding best practices in cable management, fostering regional expertise.

The International Cable Protection Committee provides essential resources on effective cable management strategies.

In summary, infrastructure development and sharing are paramount for achieving widespread, affordable broadband access throughout Africa. By coordinating efforts in national broadband expansion, enhancing cross-border connectivity, and promoting shared international infrastructure projects, the Pan-African Telecommunity (PAT) can significantly accelerate the continent's digital transformation.

The strategies and initiatives outlined in this chapter demand considerable investment, political commitment, and regional cooperation. Nonetheless, the potential benefits—ranging from economic growth and job creation to improved public services—underscore the critical importance of these endeavors for Africa's future prosperity. As the PAT moves forward with these objectives, it must remain agile, adapting to technological advancements and evolving market dynamics. Regular evaluations and adaptations of strategies will be vital to ensure that Africa's telecommunications infrastructure keeps pace with global developments and effectively meets the diverse needs of its people and economies.

CHAPTER FIFTEEN: REGULATORY HARMONIZATION OF THE PROPOSED PAT

In Africa's rapidly evolving telecommunications landscape, regulatory harmonization is a cornerstone for establishing a cohesive and efficient ecosystem. This chapter delves into the significance of creating a common regulatory framework, addresses the challenges surrounding licensing and taxation, and explores strategies to ensure fair competition and market access for all stakeholders.

15.1 Creating a Common Regulatory Framework

A unified regulatory framework is vital for promoting a seamless telecommunications market across the continent. The proposed Pan-African Telecommunity (PAT) should prioritize the following initiatives:

15.1.1 Model Regulations
1. Development of Model Regulations: The PAT should spearhead the creation of model regulations that member states can adopt or adapt to fit their unique contexts. These regulations must encompass critical areas such as licensing, interconnection, spectrum management, and consumer protection.
2. Benchmarking Against Successful Models: Drawing inspiration from the European Union's

regulatory framework for electronic communications can provide valuable insights for fostering regional harmonization.

15.1.2 Regulatory Impact Assessment
1. Establishment of Guidelines: Implementing guidelines for conducting regulatory impact assessments is crucial for promoting evidence-based policymaking across member states.
2. Leveraging International Frameworks: Utilizing frameworks such as the OECD's Regulatory Impact Analysis can serve as a guide for effective assessment practices.

15.1.3 Capacity Building for Regulators
1. Training Programs for Regulatory Authorities: The PAT should prioritize the development of comprehensive training programs tailored for national regulatory authorities, enhancing their capacity to implement effective policies.
2. Facilitating Knowledge Exchange: Creating platforms for knowledge sharing and best practice exchanges among regulators can foster collaboration and improve regulatory outcomes.

15.1.4 Regulatory Sandboxes
1. Promotion of Regulatory Sandboxes: Encouraging the use of regulatory sandboxes will enable the testing of innovative services and business models in a controlled environment.
2. Guidelines for Implementation: Developing clear guidelines for implementing sandboxes

across various national contexts will ensure their effectiveness and adaptability.

15.2 Addressing Licensing and Taxation Issues

The harmonization of licensing and taxation regimes is essential for stimulating investment and facilitating cross-border services:

15.2.1 Harmonized Licensing Framework
1. Simplification of Market Entry: The PAT should develop a harmonized licensing framework that streamlines the market entry process across multiple countries, reducing bureaucratic hurdles.
2. Technology and Service Neutrality: Encouraging technology and service neutrality in licensing will foster innovation and competition.

15.2.2 Regional Licensing
1. Feasibility of Regional Licenses: Exploring the potential for regional licenses for specific services can simplify operations for operators across borders.
2. Revenue Sharing Mechanisms: Establishing mechanisms for equitable revenue sharing among countries that participate in regional licensing will promote collaboration and mutual benefit.

15.2.3 Tax Harmonization
1. Development of Guidelines: The PAT should create guidelines aimed at harmonizing sector-specific taxes and fees, ensuring a fair and

predictable tax environment for telecommunications operators.
2. Promotion of Investment-Friendly Policies: Advocating for tax policies that encourage investment and lower service costs is crucial for enhancing digital inclusion.

15.2.4 Cross-Border Services
1. Framework for Taxation: Developing comprehensive frameworks for taxing cross-border digital services will enhance clarity and compliance.
2. Alignment with International Standards: Engaging with international initiatives, such as the OECD/G20 Inclusive Framework on Base Erosion and Profit Shifting (BEPS), will strengthen regional tax policies.

15.3 Ensuring Fair Competition and Market Access

Creating an environment conducive to fair competition and open market access is essential for fostering innovation and delivering affordable services:

15.3.1 Competition Policy
1. Guidelines for Market Assessment: The PAT should develop robust guidelines for assessing market power within the telecommunications sector, addressing potential monopolistic behaviors.
2. Frameworks for Addressing Anti-competitive Practices: Establishing frameworks to mitigate anti-competitive practices will protect

consumers and promote a healthy competitive environment.

15.3.2 Infrastructure Sharing and Open Access
1. Promotion of Infrastructure Sharing: Developing policies that encourage infrastructure sharing and open access to essential facilities will maximize resource efficiency.
2. Ensuring Non-discriminatory Access: Implementing policies that guarantee non-discriminatory access to critical infrastructure will foster a level playing field.

15.3.3 Number Portability
1. Implementation of Number Portability: The PAT should promote the implementation of number portability across member states, enhancing consumer choice and satisfaction.
2. Efficient Porting Processes: Developing standardized guidelines for efficient number porting processes will facilitate smoother transitions for consumers.

15.3.4 Interoperability Standards
1. Development of Interoperability Standards: Establishing interoperability standards for essential services, such as mobile money and the Internet of Things (IoT), will enhance user experience and integration.
2. Promotion of Open Standards: Advocating for the adoption of open standards will help prevent vendor lock-in and encourage competition.

15.3.5 Foreign Ownership Restrictions
1. Guidelines for Foreign Ownership: The PAT should develop comprehensive guidelines for foreign ownership policies, balancing national interests with the need for foreign investment.
2. Transparency in Investment Rules: Promoting transparency in foreign investment regulations will build trust and attract more investments into the sector.

15.3.6 Merger Control
1. Assessment Guidelines for Mergers: Establishing clear guidelines for assessing telecommunications mergers and acquisitions will help maintain market integrity.
2. Coordination Among Authorities: Promoting collaboration among national competition authorities for cross-border transactions will streamline the review process.

In summary, regulatory harmonization represents a complex yet vital undertaking for establishing a unified and thriving telecommunications market throughout Africa. By focusing on the development of a common regulatory framework, addressing licensing and taxation challenges, and ensuring fair competition and market access, the African Telecommunication Union (ATU) can significantly contribute to fostering a vibrant digital ecosystem. The proposed Pan-African Telecommunity (PAT), as a key initiative under the ATU, will play a crucial role in these efforts. Achieving this harmonization will demand sustained commitment, political will, and collaborative efforts among member states. Nonetheless, the potential rewards—ranging from increased investment and improved services to accelerated digital transformation—underscore the necessity of these initiatives. As the ATU progresses toward these objectives, maintaining adaptability to technological advancements and shifting market dynamics will be essential. Regular reviews and updates of regulatory frameworks will ensure their continued relevance and efficacy in the fast-paced telecommunications sector.

CHAPTER SIXTEEN: DIGITAL INCLUSION AND SKILLS DEVELOPMENT UNDER THE PROPOSED PAT

Digital inclusion and skills development are fundamental to ensuring that all African citizens can fully participate in and benefit from the digital revolution. This chapter examines comprehensive strategies aimed at bridging the digital divide, enhancing digital literacy, and nurturing innovation and entrepreneurship within the telecommunications sector.

16.1 Bridging the Digital Divide

The digital divide poses a significant barrier in Africa, characterized by unequal access to digital technologies across urban and rural landscapes and among various socioeconomic groups. To tackle this pressing issue, the African Telecommunications Union (ATU) should prioritize the following strategies:

16.1.1 Universal Service and Access Policies:
1. Establish robust guidelines that promote effective universal service and access policies.
2. Encourage the development of innovative funding mechanisms designed to extend connectivity to underserved communities.

The International Telecommunication Union's (ITU) ICT Development Fund, with its successful funding approaches, can serve as a model for advancing digital inclusion in Africa.

16.1.2 Community Networks
1. Advocate for policies that facilitate the growth of community networks.
2. Provide technical assistance and support for community-led connectivity initiatives.

The Internet Society's initiatives on African community networks showcase successful case studies in this area, providing a beacon of hope for the future.

16.1.3 Public Access Points
1. Foster the establishment of public Internet access points in libraries, schools, and community centers.
2. Develop sustainable management guidelines for these public access facilities.

UNESCO's research on public access to information offers valuable insights into effective strategies.

16.1.4 Gender Digital Divide:
1. Initiate targeted programs to address the gender digital divide specifically.
2. Enhance women's participation in the digital economy through tailored initiatives.

The GSMA's Mobile Gender Gap Report provides important data on gender-specific barriers to digital inclusion.

16.1.5 Accessibility for Persons with Disabilities:
1. Promote policies and standards that ensure digital accessibility for all users.

2. Encourage the development of assistive technologies to facilitate access.

The ITU's guidelines on digital accessibility serve as a vital resource for stakeholders.

16.2 Promoting Digital Literacy

Digital literacy is critical for meaningful engagement in the digital economy. The ATU should focus on the following key areas to enhance digital literacy:

16.2.1 National Digital Skills Strategies
1. Encourage member states to formulate comprehensive national digital skills strategies.
2. Provide guidelines for integrating digital skills into education curricula across all levels.

The European Union's Digital Education Action Plan serves as a compelling framework for digital skills development.

16.2.2 Teacher Training
1. Create specialized programs to elevate teachers' digital competencies.
2. Promote the integration of information and communication technology (ICT) in teaching methodologies.

UNESCO's ICT Competency Framework for Teachers is an excellent resource for this purpose.

16.2.3 Online Learning Platforms:
1. Encourage the establishment of localized online learning platforms tailored to African contexts.
2. Facilitate the sharing of best practices in e-learning among member states.

The African Virtual University's experiences in online education provide valuable lessons.

16.2.4 Digital Skills for SMEs
1. Develop targeted programs aimed at enhancing digital skills among small and medium-sized enterprises (SMEs).
2. Promote the adoption of digital tools to optimize business management and e-commerce practices.

The International Trade Centre's ecomConnect initiative offers essential resources for SME digitalization.

16.2.5 Media and Information Literacy
1. Initiate programs that enhance critical thinking and information evaluation skills.
2. Formulate strategies to combat misinformation and mitigate online harms.

UNESCO's extensive work on Media and Information Literacy provides comprehensive guidance.

16.3 Supporting Innovation and Entrepreneurship in the Telecom Sector

To cultivate a vibrant telecommunications ecosystem in Africa, fostering innovation and entrepreneurship is paramount. The ATU should concentrate on the following areas:

16.3.1 Innovation Hubs and Incubators:
1. Advocate for the establishment of telecom-focused innovation hubs and incubators.

2. Facilitate knowledge sharing and collaboration among innovation hubs across the continent.

The World Bank's infoDev program offers valuable insights into nurturing tech entrepreneurship.

16.3.2 Startup-Friendly Regulations
1. Develop guidelines that promote startup-friendly regulations within the telecommunications sector.
2. Support the creation of regulatory sandboxes to enable the testing of innovative services.

The GSMA's Digital Innovation Framework provides critical guidance for establishing an enabling environment for startups.

16.3.3 Intellectual Property Protection
1. Formulate guidelines to strengthen intellectual property protection in the digital economy.
2. Facilitate knowledge sharing on best practices for patent filing and protection.

The World Intellectual Property Organization's resources on IP in Africa provide essential information.

16.3.4 Access to Finance
1. Promote the development of diverse funding mechanisms for telecom startups.
2. Encourage partnerships between startups and established telecom companies to facilitate growth.

The African Development Bank's Boost Africa initiative offers important perspectives on startup financing.

16.3.5 Digital Skills for Entrepreneurs
1. Design programs aimed at enhancing both digital and entrepreneurial skills among aspiring business leaders.
2. Promote mentorship initiatives that connect experienced entrepreneurs with emerging startups.

The International Labour Organization's research on digital skills for youth offers useful structures.

16.3.6 Regional and Global Market Access:
1. Develop initiatives to assist African telecom startups in accessing both regional and global markets.
2. Support participation of African startups in international tech events and exhibitions to enhance visibility.

The International Trade Centre's initiatives focusing on tech startups in Africa provide valuable insights into market access strategies.

In summary, digital inclusion and skills development are foundational to unlocking the full potential of Africa's digital transformation. By prioritizing strategies that bridge the digital divide, promote digital literacy, and nurture innovation and entrepreneurship, the African Telecommunication Union can play a pivotal role. However, achieving these goals will require sustained effort, strategic investment, and, most importantly, collaboration among diverse stakeholders, including governments, private enterprises, civil society, and international partners. The immense potential for economic growth, job creation, and improved quality of life underscores the urgency of these efforts for Africa's future prosperity.

CHAPTER SEVENTEEN: FUNDING AND INVESTMENT STRATEGIES FOR THE PAT

The quest for robust funding and investment strategies is paramount to the advancement of Africa's telecommunications sector. As the continent strives to enhance its digital infrastructure and bridge the digital divide, this chapter delves into various approaches to secure necessary financial resources. It covers public-private partnerships, collaborations with regional development banks and international funding sources, and innovative financing mechanisms that can propel Africa's telecommunications landscape forward.

17.1 Public-Private Partnerships (PPPs)

Public-Private Partnerships (PPPs) have emerged as a pivotal mechanism for infrastructure development. By integrating public oversight with private sector efficiency, PPPs play a crucial role in effectively mobilizing resources and fostering innovation within the telecommunications sector.

17.1.1 PPP Frameworks

1. Develop tailored PPP frameworks: Create model frameworks that address the specific needs and dynamics of the telecommunications sector.
2. Risk allocation and contract structuring: Provide comprehensive guidelines that facilitate

effective risk sharing and robust contract frameworks.

The World Bank's PPP Knowledge Lab serves as an invaluable resource, offering best practices and insights into effective PPP management.

17.1.2 Capacity Building for PPPs
1. Training programs for officials: Establish training initiatives aimed at enhancing the skills of government officials in structuring and managing PPPs.
2. Knowledge sharing: Facilitate exchange programs that highlight successful PPP projects in the telecom sector.

The African Development Bank's Africa PPP Network is a critical platform for capacity building and the sharing of best practices.

17.1.3 Project Preparation Facilities
1. Strengthen project preparation facilities: Enhance existing facilities or establish new ones to streamline the development of bankable PPP projects.
2. Collaborate with existing initiatives: Work closely with established programs like the NEPAD Infrastructure Project Preparation Facility.

The insights gained from NEPAD-IPPF's initiatives in infrastructure project preparation are invaluable.

17.1.4 Regulatory Frameworks for PPPs
1. Guidelines for regulatory support: Develop guidelines that foster an enabling environment for PPPs while safeguarding the public interest.
2. Promote transparency: Implement strategies that enhance transparency and accountability throughout the PPP process.

The OECD's Principles for Public Governance of Public-Private Partnerships provide essential references for developing effective regulatory frameworks.

17.2 Regional Development Banks and International Funding

Regional and international financial institutions play a critical role in funding expansive telecommunications projects across Africa.

17.2.1 Collaboration with Regional Development Banks
1. Strengthen partnerships: Foster collaborations with institutions like the African Development Bank and the African Export-Import Bank.
2. Joint programs for financing: Develop cooperative initiatives aimed at financing telecom infrastructure projects.

The African Development Bank's "Connecting Africa" initiative exemplifies the potential benefits of such collaborative efforts.

17.2.2 Multilateral Development Banks
1. Engagement with global institutions: Build partnerships with entities like the World Bank and the International Finance Corporation to

leverage their expertise in complex infrastructure projects.
2. Support for digital development: Utilize programs like the World Bank's Digital Economy for Africa initiative to enhance support for digital transformation across the continent.

17.2.3 Bilateral Development Agencies
1. Foster partnerships: Collaborate with bilateral agencies such as USAID, DFID, and GIZ to align with their developmental priorities.
2. Program development: Create targeted programs that resonate with the strategic objectives of these agencies.

USAID's Digital Strategy serves as a reference for fostering bilateral support for digital initiatives.

17.2.4 Climate Finance for Green Telecom
1. Access climate finance: Investigate opportunities to secure climate finance for sustainable telecommunications projects.
2. Guidelines for measuring impact: Develop comprehensive frameworks for assessing and reporting the climate impact of telecom initiatives.

The Green Climate Fund's work on ICT and climate change offers valuable case studies for green financing.

17.3 Innovative Financing Mechanisms

Innovative financing mechanisms are essential for attracting the necessary investments in Africa's telecommunications sector to close the significant funding gap.

17.3.1 Infrastructure Bonds
1. Promote infrastructure bonds: Advocate for the issuance of infrastructure bonds to fund telecom projects effectively.
2. Guidelines for structuring: Develop comprehensive frameworks for the structuring and issuance of telecom infrastructure bonds.

The International Capital Market Association provides crucial insights into best practices for infrastructure bonds.

17.3.2 Blended Finance
1. Explore blended finance models: Investigate blended finance approaches that combine public, private, and philanthropic capital for greater impact.
2. Frameworks for structuring deals: Create guidelines for developing blended finance structures within the telecom sector.

The OECD offers extensive guidance on utilizing blended finance effectively.

17.3.3 Impact Investing
1. Attract impact investors: Draw in investors focused on the social and economic benefits of telecom initiatives.
2. Metrics for impact assessment: Establish standard metrics to measure and report on the impact of telecom investments.

The Global Impact Investing Network's IRIS+ system offers essential tools for impact measurement.

17.3.4 Crowd-funding and Community Financing
1. Explore crowd-funding opportunities: Investigate crowd-funding as a viable means to support community-focused telecom projects.
2. Guidelines for community-owned infrastructure: Develop frameworks that facilitate the establishment of community-owned telecom initiatives.

The Internet Society provides exemplary case studies of successful community-led financing efforts.

17.3.5 Vendor Financing
1. Promote vendor financing models: Encourage the adoption of vendor financing to support network expansions and infrastructure development.
2. Guidelines for structuring deals: Create frameworks for structuring vendor finance agreements.

Insights from the Export-Import Bank of China's support for Huawei's expansion in Africa can inform these financing models.

17.3.6 Universal Service Funds
1. Optimize Universal Service Funds: Leverage these funds for targeted infrastructure investments that address gaps in service provision.
2. Effective fund management: Develop best practices for managing Universal Service Funds and selecting impactful projects.

The ITU's study on Universal Service Funds in Africa offers pertinent examples and case studies.

17.3.7 Spectrum Auctions and Licensing

1. Innovative auction designs: Create spectrum auction models that balance revenue generation with the need for investment incentives.
2. Coverage obligations: Consider incorporating coverage obligations within licensing agreements to promote equitable access.

Research from the GSMA on spectrum pricing in developing countries provides valuable insights for effective auction design.

In summary, the development of effective funding and investment strategies is vital for bridging Africa's digital divide and unlocking the full potential of its telecommunications sector. By leveraging public-private partnerships, engaging with regional and international financial institutions, and exploring innovative financing mechanisms, the African Telecommunication Union can significantly contribute to mobilizing the resources necessary for Africa's digital transformation. The strategies outlined in this chapter necessitate close collaboration among various stakeholders, including governments, private sector entities, financial institutions, and development partners. While the challenges in securing funding are considerable, the potential returns—economic growth, job creation, and enhanced quality of life—underscore the importance of these investments for Africa's future prosperity. As the PAT pursues these strategies, it should remain adaptable to evolving market conditions and emerging financing trends. This adaptability ensures that the strategies remain resilient and effective, even in the face of changing circumstances, providing reassurance to the audience about the robustness of the plan for Africa's telecommunications landscape.

CHAPTER EIGHTEEN: IMPLEMENTATION ROADMAP FOR REALIZING THE PAT

The successful establishment of the Pan-African Telecommunity (PAT) is not merely an ambition; it requires a meticulously structured, phased approach characterized by clear milestones and robust monitoring mechanisms. This chapter delineates a comprehensive implementation roadmap aimed at guiding the formation and operationalization of the PAT, ensuring that it is both efficient and impactful.

18.1 Phased Approach to Establishing the PAT

A phased approach is crucial for gradual development, allowing for continual learning and adaptation as the PAT evolves. We propose a four-phase implementation plan:

Phase 1: Preparatory Stage (Year 1)
Stakeholder Engagement:
1. Conduct comprehensive consultations with national governments, existing regional bodies, and industry stakeholders to ensure inclusivity and representation.
2. Build consensus on the PAT's vision, mission, and core objectives to align all stakeholders.

Legal Framework:
1. Draft the PAT charter and statutes, ensuring they reflect the collective interests of member states.
2. Secure commitments and support from member states to foster a collaborative environment.

Institutional Design:
1. Develop an organizational structure and governance framework that promotes transparency and efficiency.
2. Clearly define roles and responsibilities for key positions to facilitate effective management.

The African Union's Agenda 2063 provides a foundational framework for continental initiatives that can inform this preparatory stage.

Phase 2: Establishment and Early Operations (Years 2-3)

Ratification and Launch:
1. Secure ratification of the PAT charter by member states, creating a formal commitment to the initiative.
2. Organize a high-profile launch event to raise awareness and support for the PAT.

Staffing and Operations:
1. Recruit key personnel and establish the secretariat to facilitate day-to-day operations.
2. Set up essential operational systems and procedures to ensure seamless functionality.

Initial Programs:
Launch pilot projects focusing on critical areas such as regulatory harmonization and spectrum management. The International Telecommunication Union's operational framework offers valuable insights for establishing international telecom organizations.

Phase 3: Expansion and Consolidation (Years 4-5)

Full-scale Programs:
1. Expand programs to comprehensively cover all core areas of the African Telecommunications Union's mandate.
2. Establish working groups and committees to address specialized topics and challenges.

Partnerships:
1. Formalize partnerships with regional and international organizations to enhance the PAT's credibility and resource base.
2. Engage actively with industry associations and academic institutions to foster collaboration and innovation.

Resource Mobilization:
1. Implement sustainable funding mechanisms to ensure the long-term viability of the PAT.
2. Secure long-term commitments from member states and partners to support ongoing initiatives.

The GSMA's collaboration with regional telecom organizations exemplifies successful industry partnerships.

Phase 4: Maturity and Impact (Years 6 and Beyond)

Policy Influence:
1. Establish the PAT as the authoritative voice on telecommunications issues in Africa, influencing policy at both continental and global levels.
2. Advocate for policies that promote inclusivity and growth in the telecommunications sector.

Innovation Hub:
1. Foster innovation in emerging technologies such as 5G, IoT, and AI to propel Africa's technological advancement.
2. Support the growth of African tech startups by providing funding and resources.

Digital Transformation:
Drive comprehensive digital transformation across African economies and societies, leveraging technology for economic development. The World Bank's Digital Economy for Africa initiative offers a visionary framework for this transformative phase.

18.2 Key Milestones and Timelines

Specific milestones and timelines will guide the implementation process, ensuring accountability and progress tracking:

Year 1:
1. Q2: Complete stakeholder consultations to gather diverse perspectives.
2. Q3: Finalize the PAT charter and statutes, reflecting stakeholder input.
3. Q4: Secure commitments from at least 15 member states to establish a strong foundation.

Year 2:
1. Q1: Achieve ratification of the PAT charter by initial member states.
2. Q2: Officially launch the PAT to signal its readiness to stakeholders.
3. Q3: Appoint the Secretary-General and key leadership positions to drive operations.
4. Q4: Establish the secretariat and implement basic operational systems.

Year 3:
1. Q2: Launch pilot projects focused on regulatory harmonization and spectrum management.
2. Q4: Hold the first PAT summit and industry forum to foster collaboration.

Year 4:
1. Q2: Establish all core working groups and committees to address various telecom challenges.
2. Q4: Publish the first comprehensive report on the state of telecommunications in Africa to provide insights and guide future actions.

Year 5:
1. Q2: Launch the PAT innovation fund aimed at supporting African tech startups.
2. Q4: Adopt a continent-wide framework for 5G deployment to standardize and streamline processes.

The African Union's implementation framework for Agenda 2063 serves as a model for setting continental milestones and benchmarks.

18.3 Monitoring and Evaluation Mechanisms

Robust monitoring and evaluation (M&E) mechanisms are essential for tracking progress and ensuring the effectiveness of the PAT:

18.3.1 Results-Based Management:
1. Develop a comprehensive results framework with clear, measurable indicators for each strategic objective.
2. Implement a results-based management system across all PAT programs and projects to foster accountability.

The United Nations Development Programme's handbook on planning, monitoring, and evaluating development results provides valuable guidance.

18.3.2 Annual Performance Reviews:
1. Conduct annual reviews of the PAT's performance against established targets to assess progress and identify areas for improvement.
2. Publish annual reports detailing achievements, challenges, and financial performance to maintain transparency.

18.3.3 Independent Evaluations:
1. Commission-independent mid-term and end-term evaluations are used to assess the PAT's performance comprehensively.
2. Engage external experts to evaluate the impact and effectiveness of key programs, ensuring objectivity in assessment.

The World Bank's Independent Evaluation Group offers a model for rigorous evaluation of development initiatives.

18.3.4 Stakeholder Feedback Mechanisms
1. Implement regular surveys to gather feedback from member states, industry stakeholders, and beneficiaries on the PAT's performance and impact.
2. Establish an online platform for continuous stakeholder feedback, fostering a culture of openness and responsiveness.

18.3.5 Data-Driven Decision Making
1. Develop a robust data collection and analysis system to inform decision-making processes.
2. Utilize data analytics to guide strategic decisions and adjust programs based on emerging trends and challenges.

The International Telecommunication Union's ICT Development Index methodology provides insights into telecom-sector-specific indicators.

18.3.6 Adaptive Management
1. Implement a system for regular review and adjustment of strategies based on M&E findings, promoting flexibility and responsiveness.
2. Foster a culture of learning and continuous improvement within the PAT, encouraging innovation and adaptability.

The Global Partnership for Effective Development Co-operation's work on adaptive management offers useful frameworks for enhancing organizational effectiveness.

In summary, the implementation roadmap outlined in this chapter provides a structured, strategic approach to establishing and operationalizing the African Telecommunication Union. By adhering to a phased implementation strategy, setting clear milestones, and implementing robust monitoring and evaluation mechanisms, the PAT can effectively navigate the complexities of its formation and growth. As the PAT embarks on this transformative journey, it is crucial to maintain flexibility and adaptability. The telecommunications landscape is rapidly evolving, necessitating that the PAT remains prepared to adjust its strategies and priorities in response to new technologies, market dynamics, and stakeholder needs.

The successful implementation of this roadmap hinges on strong political will, sustained commitment from member states, and effective collaboration with a diverse range of stakeholders. Ultimately, the potential impact of a well-functioning PAT on Africa's digital transformation makes this endeavor not only necessary but also profoundly worthwhile.

CHAPTER NINETEEN: POTENTIAL CHALLENGES AND MITIGATION STRATEGIES

The establishment and operation of the Pan-African Telecommunity (PAT) will inevitably encounter various challenges that could hinder its successful implementation. This chapter identifies potential barriers and proposes effective strategies to mitigate these obstacles, ensuring that PAT can fulfil its mission.

19.1 Political and Economic Barriers and Mitigation Strategies

19.1.1 Sovereignty Concerns

Many nations may be hesitant to relinquish authority to a supranational entity. To address this concern, it is essential to emphasize the principle of subsidiarity, whereby the PAT only intervenes in matters that cannot be adequately resolved at the national level. Additionally, implementing a phased approach to integration can help countries gradually increase their commitment. Insights from the African Union's experiences in managing sovereignty concerns, which have proven invaluable in navigating this issue, will continue to guide us in the right direction.

19.1.2 Diverse Policy Priorities

Member states may present conflicting policy priorities and development agendas, complicating collaboration. A flexible framework that allows for varying speeds of integration is essential. Establishing mechanisms for continuous dialogue and negotiation among member states is crucial for ongoing alignment. The OECD's work on policy coherence for sustainable development guides how to harmonize diverse national interests.

19.1.3 Economic Disparities

Significant economic disparities among member states can lead to uneven participation and benefits within the PAT. To address this, it is vital to implement solidarity mechanisms that support less developed member states, ensuring equitable participation. Programs aimed specifically at reducing digital divides within and between countries will also be crucial. The European Union's cohesion policy serves as a pertinent model for addressing regional economic disparities.

19.1.4 Funding Constraints

Securing sustainable funding for the ATU's operations and initiatives is a critical challenge. To mitigate this, diversifying funding sources—including contributions from member states, donor support, and innovative financing mechanisms—is necessary. Implementing a results-based funding model will help demonstrate the value of investments made. The African Development Bank's strategies for resource mobilization can provide relevant insights for sustainable funding practices.

19.2 Technical and Operational Hurdles

19.2.1 Interoperability Challenges

A significant challenge lies in ensuring seamless interoperability among diverse national telecom systems. To tackle this issue, it is imperative to develop and promote common technical standards across the continent. Establishing a dedicated working group focused on interoperability and standards harmonization will facilitate cooperation. The ITU's standardization initiatives provide a valuable framework for addressing interoperability issues.

19.2.2 Spectrum Coordination

Managing cross-border spectrum allocation and interference presents another operational challenge. Implementing a continent-wide spectrum management system is essential. Additionally, establishing clear protocols for cross-border coordination and dispute resolution will mitigate potential conflicts. The European Conference of Postal and Telecommunications Administrations (CEPT) provides a robust model for regional spectrum coordination.

19.2.3 Cybersecurity Threats

As telecom networks become more interconnected, so do the risks associated with cybersecurity threats. To address these risks, a comprehensive African cybersecurity framework must be developed. Establishing a continent-wide Computer Emergency Response Team (CERT) will further enhance the region's cybersecurity posture. The African Union's

Convention on Cyber Security and Personal Data Protection lays a solid foundation for these efforts.

19.2.4 Skills and Capacity Gaps

Many member states face significant technical and managerial capacity gaps. To counter this challenge, comprehensive capacity-building programs are essential. Facilitating knowledge transfer through partnerships with global telecom organizations and academic institutions will enhance local expertise. The ITU's capacity-building initiatives offer valuable resources and frameworks for strengthening skills.

19.3 Strategies for Building Consensus and Overcoming Obstacles

19.3.1 Inclusive Stakeholder Engagement

1. Multi-Stakeholder Approach: Involve governments, the private sector, civil society, and academia in the decision-making processes.
2. Regular Forums: Establish forums for ongoing dialogue and collaboration among all stakeholders.

The Internet Governance Forum's multi-stakeholder model provides a successful template for inclusive engagement.

19.3.2 Evidence-Based Policy Making

1. Research and Assessments: Conduct thorough research and impact assessments to inform policy decisions.

2. African Telecom Research Network: Establish this network to support evidence-based policymaking.

The World Bank's Development Impact Evaluation (DIME) initiative serves as an effective model for promoting evidence-based policies.

19.3.3 Phased Implementation and Quick Wins
1. Phased Approach: Start with less controversial areas to build momentum.
2. Quick Wins: Identify and prioritize projects that can deliver early value and success.

The East African Community's experiences in regional integration provide valuable lessons for phased implementation strategies.

19.3.4 Effective Communication and Transparency

Develop a comprehensive communication strategy to keep all stakeholders informed and engaged. Implement transparency mechanisms, including regular reporting and open data initiatives, to foster trust. The Open Government Partnership offers frameworks for enhancing transparency in public institutions.

19.3.5 Flexibility and Adaptability
Incorporate flexibility into the ATU's structure and processes to respond to changing circumstances. Implement regular review mechanisms to assess and adjust strategies. The OECD's work on adaptive governance offers insights into building resilient and responsive institutions.

19.3.6 Leveraging Existing Initiatives
1. Align PAT Initiatives: Ensure that PAT initiatives are aligned with existing continental programs, such as the African Continental Free Trade Area (AfCFTA).
2. Partnerships: Foster partnerships with regional economic communities and other relevant organizations to strengthen collaborative efforts.

The African Union's Agenda 2063 provides a comprehensive framework for aligning continental initiatives.

In summary, while the establishment and operation of the African Telecommunication Union will face significant challenges, these hurdles are not insurmountable. By proactively identifying potential barriers and implementing targeted mitigation strategies, the PAT can navigate obstacles effectively. The key to success lies in fostering a flexible and adaptive approach, promoting inclusive stakeholder engagement, and demonstrating clear value to member states and citizens. By addressing political and economic barriers, overcoming technical and operational challenges, and building consensus among diverse stakeholders, the PAT can play a transformative role in shaping Africa's digital future. As the telecommunications landscape continues to evolve rapidly, the PAT must remain vigilant and responsive to emerging challenges and opportunities. Regular assessment and strategic adjustment will be crucial to ensure the PAT remains effective and relevant in driving Africa's digital transformation.

CHAPTER TWENTY: CASE STUDIES AND BEST PRACTICES

This chapter examines successful regional integration efforts and exemplary global practices in telecommunications, extracting valuable lessons that can be tailored to the African context. Through a comprehensive analysis of these case studies, we aim to identify effective strategies and methodologies that could significantly benefit the proposed Pan-African Telecommunity (PAT).

20.1 Lessons from Successful Regional Integration Efforts

20.1.1 European Conference of Postal and Telecommunications Administrations (CEPT)

Established in 1959, the CEPT has been pivotal in harmonizing telecommunications policies across Europe, serving as a blueprint for collaboration.

Key Lessons:

The CEPT's adaptable framework, which allows varying levels of engagement from member states, is a beacon of hope for the African context, accommodating their diverse needs and showing that change is possible.

1. Technical Harmonization: A concentrated effort on aligning technical standards and coordinating policies enhances operational efficiency.

2. Spectrum Management: CEPT's successful cross-border spectrum allocation offers a robust model for managing shared resources.

The CEPT's Committee for ITU Policy serves as an excellent example of how to coordinate continental positions in global forums effectively.

20.1.2 ASEAN Telecommunications Regulators' Council (ATRC)

The ATRC has successfully promoted cooperation among Southeast Asian nations, fostering a collaborative telecom environment.

Key Lessons:

1. Sustained Political Momentum: Regular high-level meetings are essential to maintain focus and commitment among member states. The ATRC's commitment to actionable cooperation initiatives, such as reducing mobile roaming charges, is a testament to the tangible benefits that can be achieved through practical cooperation.
2. Capacity Building: Ongoing training programs for member state regulators empower them to meet regional challenges effectively. The ASEAN ICT Masterplan 2020 offers insightful strategies for regional digital development, showcasing a roadmap for collaborative growth.

20.1.3 West Africa Telecommunications Regulators Assembly (WATRA)

WATRA has made notable strides in harmonizing telecommunications policies across West Africa.

Key Lessons:
1. Model Regulations: The development of adaptable model regulations facilitates smoother adoption by member states.
2. Cross-Border Connectivity: Focusing on initiatives that enhance cross-border connectivity is vital for regional integration.
3. Number Portability: WATRA's successful implementation of number portability showcases the potential for harmonization.

The assembly's guidelines on quality-of-service regulation set a benchmark for establishing regional standards.

20.1.4 Eastern Caribbean Telecommunications Authority (ECTEL)

ECTEL represents a unique paradigm of a fully integrated regional telecom regulator, exemplifying how small island nations can collaborate effectively.

Key Lessons:
1. Shared Regulatory Framework: A unified regulatory approach allows multiple small states to pool resources and expertise.
2. Resource Pooling: Jointly managing limited financial and human resources maximizes operational efficiency.
3. Regional Licensing Regime: The establishment of a cohesive regional licensing framework simplifies processes for telecom operators.

ECTEL's innovative approach to spectrum management offers invaluable insights, especially in resource-constrained settings.

20.2 Adapting Global Best Practices to the African Context

20.2.1 Spectrum Management and Allocation

The U.S. Federal Communications Commission (FCC) has set the standard for innovative spectrum management practices that can be adapted for Africa.

Adaptable Practices:
1. Spectrum Auctions: Implementing auction systems can facilitate efficient spectrum allocation while ensuring fair competition.
2. Dynamic Spectrum Sharing: Embracing cutting-edge technologies that allow dynamic sharing of spectrum can enhance resource efficiency.
3. Unlicensed Spectrum: Designating specific spectrum bands for unlicensed use fosters innovation and increases access.

While adopting these auction mechanisms, the PAT should prioritize balancing revenue generation with affordability and extensive coverage objectives, sparking excitement about the innovation potential. Insights from the GSMA's research on spectrum pricing in developing nations can further guide this initiative.

20.2.2 Digital Inclusion Initiatives

South Korea's successful digital inclusion policies present a compelling framework for bridging the digital divide across diverse socio-economic landscapes.

Adaptable Practices:
1. Digital Literacy Programs: National initiatives aimed at enhancing digital literacy equip citizens with essential skills for the digital age.
2. Subsidized Access: Implementing programs that provide low-income groups with affordable devices and internet access can dramatically increase participation.
3. Public-Private Partnerships: Collaborations between public entities and private sectors in digital skills training can foster a more digitally savvy populace.

The Informatization Promotion Fund in South Korea exemplifies effective funding mechanisms for digital inclusion, which the PAT should adapt to meet Africa's unique challenges. The World Bank's Digital Economy for Africa initiative also provides a comprehensive framework for fostering digital development.

20.2.3 Regulatory Sandboxes

The concept of regulatory sandboxes, first pioneered by the UK's Financial Conduct Authority, has proven beneficial in various sectors, including telecommunications.

Adaptable Practices:
1. Innovation Testing: Creating designated environments for testing innovative services and business models fosters experimentation without undue risk.
2. Consumer Protection: Ensuring that innovation does not come at the cost of consumer safety and rights is paramount.

3. Regulatory Insights: Utilizing insights gained from sandbox experiments can help shape broader regulatory frameworks.

The ITU's guidelines on regulatory sandboxes provide a tailored framework that the PAT could adopt to stimulate innovation while managing potential risks. Relevant case studies from the GSMA on regulatory sandboxes in digital financial services can inform this approach.

20.2.4 Infrastructure Sharing

India's successful telecom tower-sharing model highlights an effective strategy for reducing costs and expanding network coverage.

Adaptable Practices:

1. Encouraging Infrastructure Sharing: Policies that promote both passive and active infrastructure sharing can significantly enhance network efficiency.
2. Independent Tower Companies: Establishing independent entities to manage shared infrastructure reduces redundancy and lowers costs.
3. Universal Service Funds: Utilizing funds to support shared infrastructure initiatives in underserved areas can ensure wider access.

Recommendations from the Telecom Regulatory Authority of India provide practical insights that the ATU can leverage to facilitate infrastructure sharing. The World Bank's studies on infrastructure sharing in emerging markets offer further guidance.

20.2.5 Cybersecurity Frameworks

The European Union's Network and Information Security (NIS) Directive outlines a robust framework for addressing cybersecurity challenges.

Adaptable Practices:
1. National Cybersecurity Agencies: Establishing dedicated agencies to oversee cybersecurity measures strengthens national defenses.
2. Incident Reporting: Mandating incident reporting for critical infrastructure operators ensures transparency and rapid response to threats.
3. Cross-Border Cooperation: Promoting collaboration between nations can enhance collective resilience against cyber threats.

The resources provided by the European Union Agency for Cybersecurity (ENISA) on implementing the NIS Directive can serve as a foundation for developing cybersecurity strategies within the African context. The African Union's Convention on Cyber Security and Personal Data Protection offers an initial framework for advancing these efforts.

The case studies and best practices explored in this chapter yield invaluable insights for the development of the Pan-African Telecommunity. While these examples provide critical guidance, it is essential to adapt them thoughtfully to align with Africa's distinctive challenges and opportunities. Key takeaways include the necessity for flexible structures that accommodate diverse member state engagements, the significance of prioritizing practical cooperation initiatives, and the importance of comprehensive approaches to digital development that transcend conventional telecom regulations.

As the PAT formulates its strategies and programs, ongoing monitoring of global best practices and emerging trends in telecom regulation and digital development will be crucial. By integrating these insights with a profound understanding of African realities, the PAT can craft innovative and contextually relevant solutions that drive the continent's digital transformation forward.

CHAPTER TWENTY-ONE: FUTURE PROSPECTS AND LONG-TERM VISION

This chapter explores the future landscape of telecommunications in Africa, highlighting the transformative role of emerging technologies, strategies for positioning the continent within the global digital economy, and the potential contributions to sustainable development goals. The Pan-African Telecommunity (PAT) stands ready to play a pivotal role in shaping this promising future, instilling a sense of confidence and reassurance in its leadership.

21.1 The Role of Emerging Technologies (5G, IoT, AI)

21.1.1 5G Networks

The imminent deployment of 5G networks across Africa is set to revolutionize connectivity, ushering in a wave of new services and applications that can significantly enhance everyday life and open up new economic opportunities.

Key Considerations:
1. Developing a Continent-wide 5G Roadmap: Establish a strategic framework to guide the deployment and implementation of 5G technology across diverse regions.
2. Addressing Spectrum Allocation Challenges: Tackle the complexities of spectrum allocation

to ensure efficient and equitable access for all stakeholders.
3. Promoting Infrastructure Sharing: Encourage collaborative approaches to infrastructure sharing among operators, reducing deployment costs and maximizing resources.

The GSMA's report on 5G in Sub-Saharan Africa illuminates the potential impacts and challenges associated with the rollout of 5G technology.

21.1.2 Internet of Things (IoT)

The Internet of Things (IoT) presents an unprecedented opportunity to revolutionize sectors such as agriculture, healthcare, and urban management, thereby significantly enhancing efficiency and productivity.

Key Considerations:
1. It is developing IoT-Specific Regulatory Frameworks: Craft regulations that specifically address the unique challenges and opportunities presented by IoT technologies.
2. Addressing Security and Privacy Concerns: Prioritize the development of robust security measures to safeguard data privacy in IoT deployments.
3. Promoting Local IoT Innovation and Entrepreneurship: Foster an environment conducive to local innovation, enabling homegrown solutions to thrive.

The ITU's "Internet of Things in Africa" report offers a comprehensive overview of the opportunities and challenges posed by IoT in the African context.

21.1.3 Artificial Intelligence (AI)

Artificial intelligence has the potential to catalyze innovation across various sectors, ranging from personalized healthcare to smart agriculture, reshaping industries and significantly improving quality of life.

Key Considerations:
1. Developing Ethical AI Frameworks: Establish guidelines to ensure the responsible and ethical development of AI technologies.
2. Promoting AI Skills Development: Invest in education and training programs to cultivate a skilled workforce capable of leveraging AI.
3. Addressing Potential Job Displacement: Proactively manage AI's socio-economic impacts by exploring reskilling and upskilling opportunities for affected workers.

The African Union's "Artificial Intelligence Strategy for Africa" serves as a roadmap for the continent's AI development.

21.2 Positioning Africa in the Global Digital Economy

21.2.1 Digital Services Export

Africa is on the cusp of a digital revolution, poised to emerge as a significant exporter of digital services. With its unique capabilities and resources, the continent is set to thrive in the global digital marketplace, inspiring optimism and a sense of pride in its potential.

1. Key Strategies:
2. Developing Digital Skills Pipelines: Invest in educational initiatives to cultivate a workforce adept in digital skills.
3. Creating Enabling Environments for Tech Startups: Foster entrepreneurial ecosystems that support innovation and the growth of tech startups.
4. Promoting African Digital Content Globally: Encourage the creation and distribution of African digital content to reach global audiences.

The World Bank's "Digital Economy for Africa" initiative offers insights into cultivating competitive digital economies across the continent.

21.2.2 Data Centers and Cloud Services

Africa, with its unique advantages, has the potential to become a significant player in the global data center and cloud services market. This potential opens up a world of opportunities and instills a sense of hope and optimism for the continent's technological advancement.

Key Considerations:
1. Developing Policies to Attract Data Center Investments: Create favorable conditions for attracting investments in data center infrastructure.
2. Addressing Energy and Connectivity Challenges: Tackle the energy and connectivity issues that could impede growth in this sector.
3. Promoting Local Cloud Service Providers: To enhance service availability, and encourage the

establishment and growth of local cloud service companies.

Reports from the African Data Centres Association provide valuable insights into the evolving African data center market.

21.2.3 Digital Financial Services

Africa is poised to lead in mobile money and digital financial innovations, creating new opportunities for financial inclusion.

Key Strategies:
1. Promoting Interoperability of Mobile Money Systems: Enhance the integration of various mobile money platforms to facilitate seamless transactions.
2. Developing Regulatory Frameworks for Fintech Innovations: Establish clear regulations that encourage innovation while ensuring consumer protection.
3. Addressing Cybersecurity Challenges in Digital Finance: Strengthen cybersecurity measures to protect users and build trust in digital financial systems.

The Alliance for Financial Inclusion's "Digital Financial Services in Africa" report provides a comprehensive overview of the sector's landscape.

21.2.4 E-Commerce and Digital Trade

The African Continental Free Trade Area (AfCFTA) presents a unique opportunity for digital trade growth, fostering intra-African commerce and access to global markets.

Key Considerations:
1. Developing Harmonized E-Commerce Regulations: Create consistent regulatory frameworks that facilitate cross-border e-commerce.
2. Addressing Cross-Border Payment Challenges: Simplify payment processes to enable smooth transactions between countries.
3. Promoting Digital Logistics Solutions: Leverage technology to enhance logistics and supply chain management in e-commerce.

The United Nations Conference on Trade and Development's (UNCTAD) "eTrade for All" initiative provides resources for promoting e-commerce development.

21.3 Contribution to Sustainable Development Goals

The ATU's initiatives can significantly contribute to several Sustainable Development Goals (SDGs), promoting holistic growth and development across the continent.

21.3.1 SDG 9: Industry, Innovation, and Infrastructure
1. Expanding Broadband Access: Improve connectivity to bridge digital divides.
2. Promoting Digital Innovation Hubs: Foster environments where innovation can flourish.
3. Supporting Tech Entrepreneurship: Encourage the growth of tech startups.

The ITU's report on "ICT-centric economic growth, innovation, and job creation" underscores the importance of ICT in achieving SDG 9.

21.3.2 SDG 4: Quality Education
1. Supporting E-Learning Initiatives: Enhance access to education through online platforms.
2. Promoting Digital Literacy Programs: Equip individuals with essential digital skills.
3. Facilitating Access to Global Educational Resources: Ensure that all learners can access quality educational materials.

UNESCO's report on leveraging ICT for education outlines strategies to enhance learning outcomes through technology.

21.3.3 SDG 3: Good Health and Well-being
1. Supporting E-Health and Telemedicine Initiatives: Improve healthcare access through digital solutions.
2. Promoting Health-Focused IoT and AI Applications: Leverage technology to enhance health outcomes.
3. Facilitating Health Information Systems: Streamline data management for better healthcare delivery.

The World Health Organization's "Global Strategy on Digital Health 2020-2025" provides a framework for advancing digital health initiatives.

21.3.4 SDG 11: Sustainable Cities and Communities

1. Supporting Smart City Initiatives: Implement technology-driven solutions for urban management.
2. Promoting IoT for Urban Management: Utilize IoT to enhance the efficiency of urban services.
3. Facilitating Digital Civic Engagement Platforms: Empower citizens through digital participation in governance.

The United Nations Human Settlements Programme's report on "ICT, Urban Governance, and Youth" highlights the role of ICT in sustainable urban development.

21.3.5 SDG 13: Climate Action

1. Promoting Green Telecom Infrastructure: Encourage sustainable practices in telecom development.
2. Supporting ICT-Based Climate Monitoring Systems: Utilize technology to track and mitigate climate change impacts.
3. Facilitating Climate-Smart Agriculture through IoT and AI: Enhance agricultural practices with sustainable technologies.

The Global e-Sustainability Initiative's "SMARTer2030" report emphasizes the potential of ICT in addressing climate change challenges.

21.4 Long-term Vision

The PAT's long-term vision should focus on positioning Africa as a global leader in digital innovation and inclusive connectivity. Key elements of this vision include:

1. Universal Broadband Access: By 2030, all Africans should have affordable, high-quality broadband access, ensuring that no one is left behind.
2. Digital Skills Leadership: Establish Africa as a premier hub for digital talent and innovation, empowering individuals with the skills needed for the future.
3. Regulatory Excellence: Develop a harmonized, forward-looking regulatory environment that fosters innovation while protecting consumer interests and promoting fair competition.
4. Digital Sovereignty: Ensure Africa's digital independence through local content development, robust data governance, and cybersecurity leadership that protects citizens' rights.
5. Sustainable Digital Transformation: Leverage digital technologies to drive sustainable development across all sectors of African economies and societies, ensuring that growth benefits everyone.

In summary, the future of telecommunications in Africa holds immense promise, with emerging technologies poised to create unprecedented opportunities for economic growth and social development. The African Telecommunication Union has a critical role to play in realizing this potential by fostering innovation, harmonizing regulations, and ensuring that the benefits of digital transformation are accessible to all Africans. As the PAT pursues its long-term vision, it must remain adaptable to the rapidly evolving technological landscape. Regularly reassessing strategies and fostering close collaboration with global partners will be essential to ensure Africa's continued progress in the digital age.

CHAPTER TWENTY-TWO: SUMMARY OF KEY FINDINGS, POLICY RECOMMENDATIONS, AND CONCLUSION

22.1 Key Findings

22.1.1 Historical Context and Current Landscape

1. Africa's telecommunications infrastructure remains fragmented, largely due to colonial-era development patterns.

2. Despite significant growth in mobile connectivity, substantial disparities persist in broadband access between urban and rural areas.

3. Regulatory frameworks vary widely across nations, hindering cross-border services and investments.

4. The digital divide continues to be a significant challenge, with many African citizens lacking access to affordable, high-quality internet services.

22.1.2 Existing Coordination Efforts

1. The African Telecommunications Union (ATU) has made progress in promoting cooperation but lacks the mandate and resources to integrate the continent's telecommunications landscape fully.

2. Regional economic communities have implemented some successful initiatives, but these remain limited in scope and geographic reach.

22.1.3 Infrastructure and Investment
1. There is a significant infrastructure gap, particularly in rural and underserved areas.
2. Current levels of investment are insufficient to meet the continent's connectivity needs.
3. Public-private partnerships have shown promise but are not being utilized to their full potential.

22.1.4 Emerging Technologies
1. 5G, IoT, and AI present significant opportunities for economic growth and social development across various sectors.
2. However, Africa risks falling behind in the adoption and development of these technologies without coordinated efforts.

22.1.5 Digital Economy
1. Africa has the potential to become a significant player in the global digital economy, particularly in areas such as digital services export, mobile financial services, and e-commerce.
2. The African Continental Free Trade Area (AfCFTA) provides a unique opportunity for digital trade growth but requires supporting digital infrastructure and harmonized regulations.

22.1.6 Sustainable Development:
1. Telecommunications and digital technologies can significantly contribute to achieving multiple Sustainable Development Goals (SDGs), particularly in education, healthcare, and climate action.
2. However, this potential is not being fully realized due to lack of coordination and insufficient investment.

22.2 Policy Recommendations
22.2.1 Establish the Pan-African Telecommunity (PAT)
1. Create a supranational body with the mandate to harmonize regulations, coordinate spectrum management, and promote infrastructure development across the continent.
2. Implement a phased approach to PAT establishment, starting with stakeholder consultations and pilot projects.

22.2.2 Harmonize Regulatory Frameworks:
1. Develop model regulations for key areas such as licensing, interconnection, and consumer protection.
2. Implement a continent-wide framework for 5G deployment and emerging technologies.

22.2.3 Coordinate Spectrum Management:
1. Establish a centralized African spectrum management system to optimize resource allocation and reduce cross-border interference.

2. Develop harmonized band plans for key services like mobile broadband and IoT.

22.2.4 Promote Infrastructure Sharing and Development:
1. Implement policies that incentivize infrastructure sharing among operators.
2. Establish a continent-wide infrastructure fund to support development in underserved areas.

22.2.5 Foster Digital Innovation and Skills Development:
1. Create a network of innovation hubs and incubators focused on telecommunications and digital technologies.
2. Develop comprehensive digital skills strategies and integrate them into national education curricula.

22.2.6 Enhance Cybersecurity and Data Protection:
1. Establish a pan-African cybersecurity framework and a continent-wide Computer Emergency Response Team (CERT).
2. Develop data protection regulations that balance innovation with privacy concerns.

22.2.7 Leverage Emerging Technologies:
1. Develop strategies for the coordinated deployment of 5G, IoT, and AI technologies across the continent.
2. Establish regulatory sandboxes to promote innovation while managing potential risks.

22.2.8 Promote Digital Financial Services:
1. Enhance interoperability of mobile money systems across borders.
2. Develop harmonized regulatory frameworks for fintech innovations.

22.2.9 Support E-Commerce and Digital Trade:
1. Implement harmonized e-commerce regulations to facilitate cross-border digital trade.
2. Address cross-border payment challenges to enable seamless transactions.

22.2.10 Sustainable and Inclusive Development:
1. Integrate telecommunications and digital technologies into strategies for achieving the SDGs.
2. Implement targeted programs to address the gender digital divide and enhance accessibility for persons with disabilities.

22.3 Conclusion

The establishment of a Pan-African Telecommunity is not just a step, but a leap towards realizing Africa's digital potential. By providing a unified framework for telecommunications development, the PAT can address the fragmentation that has historically hindered the continent's progress in this sector.

The research demonstrates that coordinated efforts in areas such as regulatory harmonization, spectrum management, and infrastructure development can yield significant benefits. These include expanded connectivity, enhanced economic opportunities, and improved delivery of essential services like education and healthcare. However, the success of this initiative will depend on sustained political commitment, adequate funding, and effective collaboration among diverse stakeholders. The challenges are significant, ranging from sovereignty concerns to technical interoperability issues, but they are not insurmountable.

The long-term vision of universal broadband access, digital skills leadership, and technological sovereignty is ambitious but achievable. By leveraging its unique strengths and embracing emerging technologies, Africa has the potential to not only bridge its internal digital divides but also to become a significant player in the global digital economy. This research provides a comprehensive roadmap for policymakers, industry leaders, and stakeholders to navigate the complex landscape of African telecommunications. Implementing the recommended strategies can position Africa for a future of inclusive growth, innovation, and sustainable development in the digital age.

The urgency of the situation demands immediate action. The proposed Pan-African Telecommunity offers a path forward, one that can transform the continent's digital landscape and unlock unprecedented opportunities for its people. By working together towards this shared vision, African nations can build a connected, prosperous, and digitally empowered future.

REFERENCES

Adelegan, O. J. (2008). Can regional cross-listings accelerate stock market development? Empirical evidence from sub-Saharan Africa. *IMF Working Paper*, 08/281.

Adelegan, O. J. (2009). The impact of the regional cross-listing of stocks on firm value in sub-Saharan Africa. *IMF Working Paper*, 09/99.

Adelegan, O. J., & Radzewicz-Bak, B. (2009). What determines bond market development in sub-Saharan Africa? *IMF Working Paper*, 09/213.

Adeola, O., & Evans, O. (2017). Financial inclusion, financial development, and economic diversification in Nigeria. *The Journal of Developing Areas*, 51(3), 1-15.

African Development Bank. (2020). The role of capital markets in economic development in Africa.

African Development Bank. (2020). Central Africa economic outlook 2020.

African Development Bank. (2020). Trade finance in Africa: Trends over the past decade and opportunities ahead. *African Development Bank Group*.

African Development Bank. (2021). Challenges and opportunities in African financial markets.

African Securities Exchanges Association (ASEA). (2020). *Annual report*.

African Securities Exchanges Association (ASEA). (2021). ASEA annual report and statistics, 2020.

African Securities Exchanges Association (ASEA). (2022). *Annual report on African stock exchanges*.

African Securities Exchanges Association (ASEA). (2023). *Annual report and statistics*.

African Securities Exchanges Association (ASEA). (2023). About ASEA. https://african-exchanges.org/about-us/

African Union. (2020). The African Continental Free Trade Area: Creating a single continental market. *AU Press*.

African Union. (2023). About the African Continental Free Trade Area (AfCFTA). https://au.int/en/cfta

African Union. (2023). Department of Economic Affairs. https://au.int/en/directorates/economic-affairs

Agénor, P. R. (2003). Benefits and costs of international financial integration: Theory and facts. *The World Economy*, 26(8), 1089-1118.

Agyei-Boapeah, H., Wang, Y., Tunyi, A., Machokoto, M., & Zhang, F. (2019). Intranational regulations, international trade, and firm value. *International Review of Financial Analysis*, 66, 101391.

Akhtar, S. (2011). Capital market integration in the ASEAN region. *Asian Development Bank Institute Working Paper Series*.

Allen, F., Carletti, E., Cull, R., Qian, J., Senbet, L., & Valenzuela, P. (2011). Improving access to financial services: The case of Kenya. In S. Edwards, S. Johnson, & D. N. Weil (Eds.), *African successes, Volume III: Modernization and development* (pp. 121-135). University of Chicago Press.

Allen, F., Carletti, E., Cull, R., Qian, J., Senbet, L., & Valenzuela, P. (2014). The African financial development and financial inclusion gaps. *Journal of African Economies*, 23(5), 614-642.

Andrianaivo, M., & Yartey, C. A. (2010). Understanding the growth of African financial markets. *African Development Review*, 22(3), 394-418.

Asongu, S. A. (2012). Government quality determinants of stock market performance in African countries. *Journal of African Business*, 13(3), 183-199.

Asongu, S. A. (2013). African stock market performance dynamics: A multidimensional convergence assessment. *Journal of African Business*, 14(3), 186-201.

Asongu, S. A. (2015). Liberalisation and financial sector competition: A critical contribution to the empirics with an African assessment. *South African Journal of Economics*, 83(3), 425-451.

Asongu, S. A., & Nwachukwu, J. C. (2018). Recent finance advances in information technology for inclusive development: A systematic review. *NETNOMICS: Economic Research and Electronic Networking*, 19(1-2), 65-93.

Baele, L., Ferrando, A., Hördahl, P., Krylova, E., & Monnet, C. (2004). Measuring financial integration in the euro area. *ECB Occasional Paper*, No. 14.

Beck, T., Senbet, L., & Simbanegavi, W. (2020). Financial sector development and integration in Africa: A review of recent developments and future prospects. *Journal of African Economies*, 29(Supplement_1), i23-i47.

Bonga-Bonga, L., & Hoveni, J. (2013). Volatility spillovers between the equity market and foreign exchange market in South Africa in the 1995-2010 period. *South African Journal of Economics*, 81(2), 260-274.

Brown, J. (2021). Challenges and opportunities in African capital market development. *Palgrave Macmillan*.

Brown, J. (2022). Cybersecurity in African capital markets: Emerging threats and best practices. *Journal of African Finance*, 18(1), 47-60.

Collier, P. (2020). African financial markets: The next frontier? *Oxford University Press*.

Demirgüç-Kunt, A., & Levine, R. (1996). Stock markets, corporate finance, and economic growth: An overview. *The World Bank Economic Review*, 10(2), 223-239.

Gourinchas, P. O., & Jeanne, O. (2006). The elusive gains from international financial integration. *The Review of Economic Studies*, 73(3), 715-741.

Hearn, B., & Piesse, J. (2010). Barriers to the development of small stock markets: A case study of Swaziland and Mozambique. *Journal of International Development*, 22(7), 1018-1037.

IMF. (2019). Regional economic outlook: Sub-Saharan Africa.

Irving, J. (2005). Regional integration of stock exchanges in Eastern and Southern Africa: Progress and prospects. *IMF Working Paper*, 05/122.

IOSCO. (2017). Objectives and principles of securities regulation. *International Organization of Securities Commissions*.

Jefferis, K. R., & Smith, G. (2005). The changing efficiency of African stock markets. *South African Journal of Economics*, 73(1), 54-67.

Koffie, A. N. B., Goyette, J., & Eloundou-Enyegue, P. M. (2021). Blockchain technology and financial inclusion in Africa. In *Extending financial inclusion in Africa* (pp. 207-226). Academic Press.

La Porta, R., Lopez-de-Silanes, F., Shleifer, A., & Vishny, R. (2000). Investor protection and corporate governance. *Journal of Financial Economics*, 58(1-2), 3-27.

Levine, R. (1997). Financial development and economic growth: Views and agenda. *Journal of Economic Literature*, 35(2), 688-726.

Levine, R., & Zervos, S. (1998). Stock markets, banks, and economic growth. *American Economic Review*, 88(3), 537-558.

Moloney, N. (2014). EU securities and financial markets regulation. *Oxford University Press*.

Muthoni, P. (2020). Equity markets and economic growth in Sub-Saharan Africa. **African Economic Research Consortium.**

Ndzamela, P. (2021). The case for an African stock exchange. *Financial Mail*. https://www.businesslive.co.za/fm/opinion/2021-03-25-phakamisa-ndzamela-the-case-for-an-african-stock-exchange/

Ntim, C. G. (2012). Why African stock markets should formally harmonise and integrate their operations. *African Review of Economics and Finance*, 4(1), 53-72.

Ntim, C. G., Opong, K. K., & Danbolt, J. (2011). The relative value relevance of shareholder versus stakeholder corporate governance disclosure policy reforms in South Africa. *Corporate Governance: An International Review*, 20(1), 84-105.

Otchere, I., & Soumaré, I. (2021). The African Continental Free Trade Area: Opportunities and challenges for capital market integration and development. In *African Economic Development* (pp. 469-493). Emerald Publishing Limited.

Pagano, M., & Padilla, A. J. (2005). Efficiency gains from the integration of stock exchanges: Lessons from the Euronext 'natural experiment'. Report for Euronext.

Quaglia, L. (2010). Completing the single market in financial services: The politics of competing advocacy coalitions. *Journal of European Public Policy*, 17(7), 1007-1023.

Rossouw, G. J. (2005). Business ethics and corporate governance in Africa. *Business & Society*, 44(1), 94-106.

Rossouw, G. J., & Stuckelberger, C. (Eds.). (2005). *Globalisation and corporate governance*. Springer.

Senbet, L., & Otchere, I. (2010). African stock markets: Challenges and opportunities. *African Development Review*, 22(3), 601-634.

Yartey, C. A. (2008). The determinants of stock market development in emerging economies: Is South Africa different? *IMF Working Paper*, 08/32.

Zhao, X. (2014). Barriers to the integration of African stock exchanges: A case study of West Africa. *African Development Review*, 26(3), 420-433.

www.ingramcontent.com/pod-product-compliance
Lightning Source LLC
Chambersburg PA
CBHW062103220526

45471CB00010B/3585